The Fundamental Processes in Ecology

The Fundamental Processes in Ecology

Life and the Earth System

Second edition

David M. Wilkinson
Visiting Professor of Ecology,
University of Lincoln, UK

OXFORD
UNIVERSITY PRESS

Great Clarendon Street, Oxford, OX2 6DP,
United Kingdom

Oxford University Press is a department of the University of Oxford.
It furthers the University's objective of excellence in research, scholarship,
and education by publishing worldwide. Oxford is a registered trade mark of
Oxford University Press in the UK and in certain other countries

First Edition published in 2006
Second Edition published in 2023

Published in the United States of America by Oxford University Press
198 Madison Avenue, New York, NY 10016, United States of America

British Library Cataloguing in Publication Data
Data available

Library of Congress Control Number: 2023938491

ISBN 978–0–19–288464–0
ISBN 978–0–19–288465–7 (pbk.)

DOI: 10.1093/oso/9780192884640.001.0001

Printed and bound by
CPI Group (UK) Ltd, Croydon, CR0 4YY

In memory of James Lovelock (1919–2022).

Preface

This book was first published in 2006 and was an expanded version of a paper published in the journal *Biological Reviews* (Wilkinson, 2003). Much has changed since the first edition came out, and the second edition is substantially revised—including a new chapter on dispersal and substantial rewriting in the final chapter. Changes in publishing have also allowed this edition to be more extensively illustrated with colour photographs (the photos are all my own). The book's central argument, that academic ecology needs to be reimagined in the context of the whole Earth system, has I think become more mainstream because of the increasingly urgent need to address a whole set of global environmental problems. This is not directly an applied ecology book; its aim is to prompt people to think hard about the underlying ecological ideas that should underpin our approach to the Earth system; and the crucial role of life in creating the environment we value but increasingly also threaten by our own actions. As I said in the preface to the first edition, I don't expect anyone to agree with all the ideas and suggestions in this book. I will consider the book successful if it provokes people into reconsidering the core ideas of ecology, and hopefully in so doing causing some of them to generate new and interesting ideas themselves.

Many people have helped with the writing of this second edition. I thank them all. Comments and suggestions on ways to structure the second edition were supplied by Tom Barker, Roberto Cazzolla Gatti, Joel Cohen, Ford Doolittle, Jane Fisher, Tim Lenton, Graeme Ruxton, Ian Sherman, and several anonymous referees for Oxford University Press. Joel Cohen suggested the new subtitle. Graeme Ruxton also commented on drafts of all the chapters, as did my wife Hannah O'Regan. Comments on draft text and/or discussion of points of detail were provided by Rudy Arthur, Karl Friston, Stuart Humphries, Tim Lenton, Jeremy Midgley, Arwen Nicholson (whose PhD work, supervised by Tim Lenton and me, also influenced the text), Sergio Rubin, and Jun Yang. James Lovelock's ideas have been a big influence on both editions of this book, and he commented on an early draft of the first edition. Jim died—at the great age of 103—during the writing of this second edition. This book is dedicated to Jim with thanks for over two decades of inspirational discussions of science.

David M. Wilkinson
Visiting Professor of Ecology
University of Lincoln
April 2023

Contents

'The answers to nearly all the major philosophical questions are either found in or illuminated by the science of life, especially ecology, whose stated goal is the elucidation of the relationship of organisms to environment.'

Lynn Margulis (in Margulis and Sagan, 1997, p 311).

'If there is any subject that deserves lucid attention, it is that of the condition of ecology in the modern world.'

Bruno Latour (2018, p 45).

PART I

Introduction

'In fact, the conditions necessary for life on the earth have not been "naturally" there but have been shaped by life itself.'

Hailia (1999a, p 338).

CHAPTER 1

Introducing the thought experiment

1.1 The entangled bank

The subject matter of ecology appears confusingly complex, and one of the most famous images of this complexity in the scientific literature is Charles Darwin's description of an entangled bank (Figure 1.1). In the closing pages of *On the Origin of Species* he wrote, 'It is interesting to contemplate an entangled bank, clothed with many plants of many kinds, with birds singing on the bushes, with various insects flitting about and worms crawling through the damp earth' (Darwin, 1859, p 489). A quarter of a century ago in an essay in the journal *Oikos* I argued that Darwin's image ignored many of the most important parts of the system, such as the microorganisms in the soil and the mycorrhizal fungi in the plant roots (Wilkinson, 1998). While Darwin stressed the animals (birds, insects, and worms), it is the plants, fungi, and, most importantly, the microbes that are involved in the majority of ecosystem services (Figures 1.2 and 1.3). Indeed, for most of the history of life on Earth ecology was entirely microbial, although today microbes are still surprisingly rare in our general ecology textbooks. In that 1998 essay I wrote 'To a first approximation the animals are of no importance to the functioning of the system'. Possibly I was slightly too hard on the animals and their lack of functional importance. As Joel Cohen (pers. comm.) has pointed out to me, while animals may not be required for most of the important ecosystem processes, they should be regarded as potentially important catalysts, as their presence may greatly speed up the *rate* of these processes. For example, Darwin's worms are probably catalysing the microbial breakdown of leaf litter.

Ecological systems such as Darwin's entangled bank (composed of myriad species of microbes, fungi, plants, and animals, all interacting with each other and their abiotic environment) are clearly difficult to understand and challenging to explain to students and the wider public. It is instructive to ask, 'How has academic ecology attempted to order this complexity?' I believe the answer to this question can be most easily seen by looking at university-level general ecology texts. As Stephen Jay Gould (2002, p 576) wrote in his *magnum opus*, 'yes, textbooks truly oversimplify their subjects, but textbooks also present the central tenets of a field without subtlety or apology—and we can grasp thereby what each generation of neophytes first imbibes as the essence of the field ... I have long felt that surveys of textbooks offer our best guide to the central convictions of an era'.

1.2 The entity approach

How do textbooks organize ecology? In most cases the answer is as a hierarchy of entities (I am indebted to Haila, 1999b for the term 'entities' in this context). There appears to be a reasonable consensus about how to classify these ecological entities in a hierarchical manner—going from genes through individuals, populations, species, communities, and ecosystems to the biosphere (e.g. Colinvaux, 1993; Krebs, 2009; Ricklefs and Relyea, 2014; Begon and Townsend, 2021) A hierarchical approach, from population to community ecology, was also used to structure the *Principles of Animal Ecology* (Allee et al., 1949), one of the key textbooks in the mid-twentieth century. This approach has an even longer pedigree in plant ecology with two

The Fundamental Processes in Ecology. Second Edition. David M. Wilkinson, Oxford University Press.
 DOI: 10.1093/oso/9780192884640.003.0001

Figure 1.1 Downe Bank nature reserve in Kent, England. This is a short walk from Down House where Darwin lived from 1842 until his death in 1882. The Bank was a favourite afternoon walk for Charles and his wife Emma and the site of some of his orchid research. It's often suggested that this was the site he had in mind when he wrote of an 'entangled bank' at the end of *On the Origin of Species*—although many of the banks and field margins around the village of Downe would have similarly matched his description in the nineteenth century (Wilkinson, 2021a).

important early texts (Schimper, 1903; Warming, 1909) using a different entity-based strategy, being arranged around various plant communities. However, Currie (2011) suggests that it was E.P. Odum in the mid-twentieth century who was key to popularizing this biotic hierarchy as an organizing principle in ecology in general, and textbooks in particular. An early exception to these approaches was Charles Elton's classic text *Animal Ecology* from the 1920s (Elton, 1927), which he organized, at least in part, around concepts such as 'succession', 'parasites', and 'dispersal'. Ricklefs (1976) also wrote a more concept-orientated introductory ecology text, but reverted to a more entity-based approach in later textbooks (e.g. Ricklefs and Relyea, 2014).

An entity-based approach has great strengths in describing systems. It is probably also inspired by a strongly reductionist tradition which believes that the lower levels in a hierarchy contain all the information needed to understand the higher levels. Reduction has been so successful in tackling many problems in the past that the philosopher Mary Midgley (2001) has argued there has been an unfortunate tendency for some scientists to think that it is not only necessary but also sufficient to explain any scientific problem. In this view, community ecology is just population ecology writ large, while an understanding of communities will provide everything required to understand the biosphere. An interesting example of this approach is given by Allen et al. (2005) who describe a 'bottom-up' model which attempts to make predictions about the global carbon cycle based on the effects of body size and temperature on individual organisms. However, such bottom-up, reductionist

Figure 1.2 A biological soil crust (cryptogamic soil or cryptogamic ground cover) in Utah Desert, USA. Such crusts often have filamentous cyanobacteria as an important part of their structure, along with mosses, lichens, and assorted microorganisms. These crusts are very important in many arid and semi-arid systems around the world, especially in controlling hydrological aspects of the soil and preventing soil erosion (Lange et al., 1992). They are thought to be similar to the earliest terrestrial communities (Mitchell et al., 2020) and provide a rare example of a system where the ecological importance of microbes is clearly visible to the unaided human eye. Mainstream ecology has tended to concentrate on macroscopic organisms (Fenchel, 1992; Wilkinson, 1998); however, there is now an increase in microbial papers in several ecological journals, although this work has yet to feature in many ecology textbooks. As an illustration of the problem, while writing the new version of this chapter (in 2022) I was also a minor author on a microbial biogeography paper which was submitted to a reasonably prominent ecology journal. The paper was rejected without review as not of wide interest—indeed the journal has something of a reputation for rejecting papers on protists as, almost by definition, not interesting or significant, although in the cases I have been told about they are often later published in other prominent journals (though not always ecology-focused ones). Microbial studies are crucial for the development of Earth systems ecology; as such they feature prominently in this book and need to have a greater profile in mainstream ecology journals (see also Figure 1.3).

approaches may have limits when faced with all the complexities of the entangled bank. In studying a complex system it is often its organization that is most important. Ernst Mayr, in his last book, illustrated this point with the following physiological example: 'No one would be able to infer the structure and function of a kidney even if given a complete catalog of all the molecules of which it is composed' (Mayr, 2004, p 72). In an ecological context, Lawton (1999) has argued that the many complex contingencies in ecological systems may limit such an approach, forcing us to rely less on reduction and experimental manipulations, especially at the level of community ecology.

1.3 A process-based approach

An obvious alternative to the hierarchical entity approach would be to emphasize processes, especially if the goal is conceptual understanding rather than a narrative description of the natural world. In the late-twentieth century various ecologists started thinking along these lines—for example see Currie (2011) for a discussion of hierarchy versus process in ecology which is independent of the arguments in the first edition of this book. This approach immediately raises a crucial question: 'What are the fundamental processes in ecology?' While many authors appear to agree on the broad outline of an entity approach (genes to biosphere), no such consensus is available for ecological processes. This book is a provisional attempt to address this difficult question.

The type of approach used is important as it can govern the kind of ecological questions a researcher asks. Considering peatland systems, the entity approach suggests questions about the number of different peatland types. Such questions date back to Linnaeus in the eighteenth century, who appears to have been the first person to publish lists of plant species from different types of bogs (DuRietz, 1957). Much more recently the British National Vegetation Classification has defined 38 peatland plant communities to be found in Great Britain (Rodwell, 1991). A process-based approach to peatlands would more naturally lead to very different questions, for example about their role in carbon sequestration and its climatic implications (e.g. Klinger et al., 1996; Clymo et al., 1998; Loisel

Figure 1.3 By definition you usually need a microscope to see microorganisms; however, sometimes their effects can be viewed without microscopy. Here, in a farm dung heap in the English Midlands on a cold winter's morning, the heat generated by microbial metabolism can be seen as steam. Large amounts of heat can be generated by microbes in dung and compost heaps; indeed the highest recorded temperature at which an animal is known to be able to complete its life cycle is 60°C for a nematode in a compost heap (Clarke, 2017). See also Table 1.1.

et al., 2021). A 2004 analysis of ecological research papers published over the preceding 25 years suggested that there was a growing increase in studies of processes (Nobis and Wohlgemuth, 2004; see also Currie, 2011). If the approach taken can affect the question asked then it can clearly affect our understanding of Earth. If a more process-based approach is to be considered then it becomes important to develop a reasonably rigorous way of defining key, or fundamental, ecological processes.

One possibility would be to ask ecologists what processes they consider important. In the run-up to the 75th anniversary of the British Ecological Society a survey of its membership asked members to rank a list of ecological concepts in order of their importance (Cherrett, 1989). The resulting top five concepts were 'the ecosystem', 'succession', 'energy flow', 'conservation of resources', and 'competition'. It is interesting that some of these overlap with those used by Elton (1927) in his early attempt at a concept-based ecology text. However, Cherrett's concepts differ from what I mean in this book by 'fundamental processes'; for example 'conservation of resources' only applies to a planet populated with intelligent organisms that can plan ahead. Other concepts discussed by Cherrett, such as the idea of nature reserve management or maximum stainable yields (by harvesting humans), are also clearly not fundamental ecological processes but important applied concepts for a planet with intelligent life. However, an understanding of process can indeed be very important in more applied contexts—as illustrated by the example of peatlands, carbon sequestration, and climate change. In this case we need to not only know how much carbon is stored in peatlands but also understand the processes that lead to this sequestration.

An alternative strategy would be to attempt to approach fundamental ecological processes experimentally using mesocosms. However, it is unclear if such systems are large enough (and they clearly operate over an unrealistically short timescale) to answer such big conceptual questions. The largest of these experimental systems has been the 1.3-ha Biosphere 2 closed-environment facility in Arizona, USA. However, arguably the main theoretical contribution of this mesocosm has been to illustrate the difficulties in maintaining stable ecological systems, even on a year-to-year basis (Cohen and Tilman, 1996). Therefore even the largest mesocosm apparently does not provide a realistic system for experimental study of many major ecological processes. It is unclear if size is the most important issue here; for example Milcu et al. (2012a) argued that a better way forward for using closed ecological systems in carbon cycling models might be to have smaller systems which would allow much more replication. As Alex Milcu and colleagues pointed out, physical models of systems have often played an important role in many areas of science, and so it's possible we have been putting too much reliance on computer models in our attempts to understand ecosystem processes such as carbon cycling (see for example Milcu et al. 2012b). An alternative approach is to use the idea of thought experiments. While these are commonly used in philosophy and theoretical physics they are less common in ecology and the environmental sciences. However, thought experiments have a long history of use in addressing problems where direct experiment is difficult or impossible (see Sorensen, 1991 for a useful short introduction). While a conventional experiment usually adds new data from the 'real world', a thought experiment sets up imaginary scenarios with the intention of adding to understanding by investigating the experiment's internal consistency or compatibility with what is already known.

Although more common in other subject areas, thought experiments are not unknown in ecology. A well-known example is W.D. Hamilton's (1971) paper *'Geometry for the selfish herd'*. This paper started thus: 'Imagine a circular lily pond. Imagine that the pond shelters a colony of frogs and a water-snake.' Hamilton famously went on to use this simplified imaginary scenario to test the logic of ideas about the anti-predator advantages of living in groups. This paper is now considered a classic of evolutionary ecology, showing that thought experiments can sometimes be powerful tools for thinking about ecological ideas. In Robert Holt's (2020, p 93) words they can help us gain 'a conceptual hook on the complexity of the world'.

This book is structured around the following thought experiment. 'For any planet with carbon-based life, which persists over geological timescales, what is the minimum set of ecological processes that must be present?' By limiting myself to considering carbon-based life located on a planet, many possible life forms are excluded. For example, the astronomer Fred Hoyle invented extraordinary intelligent interstellar clouds in his novel *The Black Cloud* (Hoyle, 1957). This is one of the most interesting alien life forms in science fiction as it relies on neither Earth-type biology nor a planetary habitat. Writing as a scientist he also argued for carbon-based life living in comets, although many have viewed this as another aspect of Hoyle's science-fiction writing. Indeed it is possible that Hoyle himself may have become less convinced by some of the more extreme versions of these ideas towards the end of his life (Gregory, 2005). Clearly one can imagine potential biological systems and ecologies that fall outside the carbon-based, planet-based boundaries of the discussion in this book. However, to be useful a thought experiment needs constraints to limit the scope of the speculation. The constraints in my thought experiment seem reasonable, as Pace (2001) has outlined arguments that suggest that any life in the universe is likely to be based on organic chemistry (Box 1.1). In addition, we have studied one example (Earth), and similar conditions may exist around many main sequence stars (Kasting et al., 1993; Franck et al., 2004; Kasting, 2010). Indeed very un-Earth-like places may allow life to survive if a heat source allows life access to water in a form that allows solute diffusion (Pace, 2001). Possible examples include heat from internal planetary sources or 'impact oases' producing liquid water on an otherwise frozen astronomical body, due to an asteroid or comet impact (O'Brien et al., 2005). While questions about the possibility of life on planets other than Earth are fascinating, my main intention with this book is to use an astrobiological perspective as a means of thinking about the fundamental processes of ecology on Earth. In the context of all the changes we are making to our planet, an attempt to recast academic ecology from such an Earth systems perspective appears worthwhile.

Box 1.1 The universal nature of biochemistry.

Norman Pace (2001) outlined a number of reasons for thinking that any life in the universe will be based on organic chemistry. Here I briefly summarize his main arguments; see his original essay for more details.

Only two atoms are known to serve as the backbone of molecules large enough to carry biological information: carbon and silicon. Carbon is much more likely because:

1) Carbon readily forms chemical bonds with a wider range of other atoms than silicon.
2) The electronic properties of carbon allow the formation of double and triple bonds with other atoms, while silicon does not readily form such bonds. These bonds allow for the capture and storage of energy from the environment, a crucial ability for any form of life.

Which organic chemicals?

Within the 'chemical space' of possible organic chemicals, life on Earth appears to use a rather limited selection. This may be an example of historical accident (in which case different organic chemicals may be used by any life elsewhere in the universe) or it may be due to chemical constraints which could apply to all life forms. For example the requirement of solubility in water may be a reason why many of the smaller organic molecules used by life on Earth are derivatives of simple carboxylic acids and organic amines (Dobson, 2004); however, water is not the only solvent for organic chemistry so there is at least an outside chance that is not necessary for carbon-based life (Ball, 2005a).

1.4 The Gaian effect

In the case of Earth, life is not a new phenomenon but has existed for much of its history. Our planet is around 4,600 million years old and many authorities consider it likely that life was present *at least* 3,500 million years ago; indeed this figure has been cited in many review articles and textbooks in recent decades (e.g. Brasier, 1979; Raven and Johnson, 1999; Lenton and Watson, 2011). The very earliest dates don't rely on conventional fossils but on the interpretation of carbon isotopes recovered from rocks around 3.8 billion years old in Greenland (Rosing, 1999; Nisbet and Sleep, 2001); these results have been very controversial, with doubts being raised over both the evidence for life and the dating of the rocks (Lepland et al., 2005; Moorbath, 2005). The isotope evidence is based on the ratio of ^{13}C to ^{12}C and is suggestive of some type of photosynthesis (although not necessarily the most familiar oxygenic photosynthesis)—clearly something of great ecological significance if correct (Lenton and Watson, 2011). The oldest fossil microorganisms are considered by many to come from the Apex chert in Australia and are in excess of 3,465 million years old (Schopf, 1999) and are responsible for the textbook dates of 3.5 billion years for the first life. However, these too have attracted controversy, having been reinterpreted as secondary artefacts of amorphous graphite formed around hydrothermal vents (Brasier et al., 2002). Whatever the nature of these 'fossils', Brasier et al. (2002) conceded that as with the Greenland rocks described above, 'carbon isotopic values from the graphite cherts imply a significant biological contribution to the carbon cycle' at the time, so even if Schopf's microfossils turn out to be artefacts there is still good circumstantial evidence for life at this date. Five hundred million years later, by around 3 billion years ago, there is abundant evidence for life, with relatively uncontroversial fossil evidence for microbial life on Earth, such as stromatolites (Figure 1.4) described from several different locations (Schopf, 1999; Nisbet, 2001; Nisbet and Sleep, 2001).

Currently it seems safe to claim secure evidence for life at 3 billion years, good circumstantial evidence at 3.5 billion, and a real possibility that life on Earth is older, although currently we are

Figure 1.4 Fossil stromatolites at Cooper's Cave, in the Cradle of Humankind (the site is also known for fossil hominins), northwest of Johannesburg in South Africa. These stromatolites are from around 2.6 billion years ago when subsidence led to the area being flooded. In these shallow seas life thrived and stromatolites were common, likely formed by photosynthetic cyanobacteria living in the well-lit shallow waters (McCarthy and Rubidge, 2005).

lacking good evidence for this (Antcliffe et al., 2017). Despite 15+ years of additional research these conclusions haven't substantially changed since the first edition of this book. Prior to 3.8 billion years ago, conditions for life on Earth would have probably been very challenging. Evidence from our Moon (where the relevant impact craters are preserved unlike the situation on Earth) has suggested that the early history of Earth would have been marked by a period characterized by large asteroid impacts (the so-called Late Heavy Bombardment), which could have made any surface life impossible; this would rule out photosynthetic organisms which by definition must live on the surface of a planet to gain access to life (Sleep et al., 1989). However, some reinterpretations of lunar data suggest caution is required when thinking about these previously widely accepted ideas (Spudis et al., 2011), and some astronomers now think that a more prolonged period of asteroid impacts (from approx. 4.2 to 3.4 billion years ago) may be more likely than an intense spike in such events around 3.9 billion years ago implied by the term 'Late Heavy Bombardment' (Zellner, 2017). This matters as if there was life present and there was indeed a period of significant asteroid impacts then we need to consider how life survived.

There appears to be two main ways in which it could have happened. The first possibility would be by living deep within Earth's rocks—the 'deep hot biosphere' popularized by Thomas Gold (1999)—see Colman et al. (2017) for a recent review of these ideas. The second possibility is life surviving off our planet, the most widely discussed possibility being the colonization, or recolonization, of Earth from Mars via microbes transported in meteorites originating from planetary surfaces (Davies, 1998). An interesting variation on this second idea is that during a large asteroid impact on Earth microorganisms could be ejected into space inside rocks, which later fall back onto the now sterile Earth, so reseeding our planet (Wells et al., 2003). Clearly these various possibilities are not mutually exclusive.

Whatever the date of the first life on Earth it is clear that it has existed for at least 3 billion years and this raises an interesting problem—given all the things that could have happened to cause its extinction, why has it survived so long? The range of potential disasters is wide, including the impact of a large meteorite (objects in excess of 500 km in diameter could potentially vaporize Earth's oceans [Nisbet and Sleep, 2001]) or global ice ages, extreme versions leading to 'Snowball Earth'. More complex life of the kind that most ecology courses concentrate on tends to be less tolerant than prokaryotes (Table 1.1 illustrates this for the upper temperature limit for growth) and as such it could be exterminated by less extreme events than those described above.

While ecology has traditionally considered the abiotic environment as an unchanging stage on

Table 1.1 Approximate upper temperature limits for growth for a range of organisms at which they can complete their life cycle (data from Clarke, 2017). A comparison with the first edition of this book (Wilkinson, 2006) shows that for several groups the maximum recorded temperature has increased over the last 20 years. At low temperatures, with the exception of vascular plants, some members of all major taxa can grow below 0°C. The lower limit shows much less variation than the upper limit, but some bacteria can grow at −20°C, and the key to this is that the physical properties of ice allow solute diffusion at temperatures below 0°C (Pace, 2001). See also the caption of Figure 1.3.

Taxon	Temperature (°C)
Archaea	122
Bacteria	100
Single cell eukaryotes	60
Metazoa	60
Vascular plants	65

which organisms act out their ecological relationships (e.g. Hutchinson, 1965) there is another fascinating possibility—that feedbacks between the organisms and their environment have helped to maintain habitable conditions on Earth. This is the Gaia hypothesis, developed by James Lovelock and Lynne Margulis in the late 1960s and 1970s. (For more recent descriptions of this idea see Lovelock, 2000a; Lovelock, 2003; Lenton and Watson, 2011; Lenton et al., 2018. For criticisms see Kirchner, 2003; Tyrrell, 2013; Waltham, 2014.) While early versions of the Gaia hypothesis described 'atmospheric homeostasis by and for the biosphere' (Lovelock and Margulis, 1974) a more modern definition is 'that organisms and their material environment evolve as a single coupled system, from which emerges the sustained self-regulation of climate and chemistry at a habitable state for whatever is the current biota' (Lovelock, 2003, p 769).

I will discuss Gaia theory in much more detail in chapter 12, although its influence is widespread throughout this book; at this point I will simply point out that it raises the interesting possibility that life may be involved in helping to maintain conditions suitable for its own existence on a planet such as Earth. In this context it is worth asking, for any of my putative fundamental processes, what would be its long-term effect on the habitability of any planet on which it occurred? I have referred to this concept as the 'Gaian effect' of the process. Any process that tends to increase the survival of life on a planet is said to have a positive Gaian effect, while one that decreases the chances of survival has a negative Gaian effect. This term has the advantage of brevity, but it is not ideal since Gaia is usually defined as the whole system on a planet; thus using it to refer to part of the system is potentially confusing (Lenton and Wilkinson, 2003). However, I have stuck with the term (used in the original paper [Wilkinson, 2003] which was the forerunner to this book) as I consider the advantages of brevity outweigh the possibility for confusion and the term has the additional merit of making the origin of the idea, from Lovelock's Gaia theory, instantly apparent.

1.5 Overview

In this book I address the question 'What are the fundamental processes in ecology?' by considering the following thought experiment: *for any planet with carbon-based life, which persists over geological timescales, what is the minimum set of ecological processes that must be present?* I have identified eight putative fundamental processes which are described in the chapters constituting Part II of this book. I have not been overly concerned with a formal definition of 'process' in this context; other scientists would no doubt split (or lump) together some of my suggested 'fundamental' processes. Their main utility in this book is as a way of organizing ecology in a novel and hopefully thought-provoking way, where the emphasis is on placing ecology in a planetary—or Earth systems—context. These processes (and their Gaian effects) are summarized in Table 1.2. Some processes, such as natural selection and competition, are characteristic of life in general rather than ecology in particular. As such they fall outside the main topic of this discussion but will be considered in the context of several of the ecological processes in the following chapters. The final part of this book will consider the whole Earth system in the context of these proposed fundamental processes.

Table 1.2 Summary of the fundamental processes of ecology. The suggested Gaian effect (+, –, or ?) is given in parentheses.

Energy flow: required by the second law of thermodynamics, leading to energy consumption and waste product excretion. Chapter 2.
Multiple guilds: autotrophs, decomposers (and possibly parasites) (+). Chapter 3.
Trade-offs: leading to within-guild **biodiversity** (+). Chapter 4.
Dispersal: required for long-term survival (+). Chapter 5.
Ecological hypercycles (?; + if efficient nutrient cycling present). Chapter 6.
Merging of organismal and ecological physiology (+ or –). Chapter 7.
Photosynthesis? (not required but possibly likely in most biospheres). Chapter 8.
Anoxygenic (+)
Oxygenic (+ or –)
Carbon sequestration (+ or –). Chapter 9.
Nutrient cycling as an emergent property of the fundamental processes (+). Chapter 10.

PART II

The Fundamental Processes

'We already have far more facts than we can handle. What we need most is to improve our ways of sorting and relating them—to work on concepts, to philosophize.'

Midgley (1989, p 53).

CHAPTER 2

Energy flow

2.1 The second law of thermodynamics

The second law of thermodynamics is so important to many areas of science that the scientist and novelist C.P. Snow (1959) famously argued that knowledge of it was a crucial test of scientific literacy during his famous Rede Lecture on 'the two cultures'. Snow thought of the Arts/Humanities as one culture and Science as the other and suggested that few people in the Humanities would be able to pass his second-law-based Science literacy test. Although it only occasionally gets a mention in ecological discussions, the second law lies at the heart of fundamental ideas about energy flow in ecosystems.

At its simplest, the second law states that 'heat flows from a hotter to a colder body' or that energy becomes increasingly dispersed—with the heat energy in the hotter body becoming dispersed across the whole system. This means that entropy (Box 2.1) is greater (or at least not lower) after any given process has taken place (Penrose, 2004). In more formal notation the second law can be expressed as the inequality:

$$dS \geq 0$$

where S stands for entropy.

Box 2.1 Entropy and information.

The principal sources for my approach in this box are Lovelock (2000a), Stewart (2012), and Kleidon (2016).

Entropy (S) is the measure of the disorder of a system and is a notoriously difficult concept. A relatively simple way to represent it is Boltzmann's equation:

$$S = k \ln P$$

where k = Boltzmann's constant (1.38×10^{-23} JK^{-1}). P = a measure of disorder (technically the number of distinct microstates in the system); effectively it's a measure of probability or the 'degree of surprise' (Aleksander, 2002).

Life is 'surprising' (there are many more ways of assembling molecules into a non-living mess than there are to make an ordered, functioning organism) and as such has low probability (P) and hence low entropy (S). More generally there are many more ways of being disorganized than organized. So unless effort is expended on staying organized, random changes (such as the jostling of molecules caused by heating) quickly lead to a disorganized state.

The idea of entropy appears closely related to Claude Shannon's concept of information. Entropy can roughly be thought of as a measure of 'randomness' in a system, while information can be thought of as something like a lack of randomness. However, if you talk to a range of physicists with an interest in biology—or biologists with a good knowledge of classical physics—you will find a wide range of views on exactly how useful the apparent resemblance between thermodynamic entropy and Shannon information actually is. Such concerns have a long history (e.g. Medawar, 1982—indeed his essay, namely '*Spence and general evolution*', was first published in 1963).

At first sight organisms appear to violate the second law with their information-rich (low entropy) nature. Indeed organisms have been described as 'islands of order in an ocean of chaos' (Margulis and Sagan, 1995). The way in which organisms manage to achieve this is well known but far from trivial, as it lies at the core of ecology. As is so often

The Fundamental Processes in Ecology. Second Edition. David M. Wilkinson, Oxford University Press.
 DOI: 10.1093/oso/9780192884640.003.0002

the case with overly familiar points, there is a real danger of overlooking them; indeed many ecology textbooks do not discuss the ideas I discuss in this chapter. Over the course of its history attempts to understand ecology through the second law have been proposed on multiple occasions and fallen in and out of academic fashion. Especially noteworthy was the work of Alfred Lotka in the early twentieth century and Howard Odum in the mid-twentieth century (Chapman et al., 2016).

2.2 Schrödinger, entropy, and free energy

In 1944 the physicist Erwin Schrödinger published a short book entitled *What Is Life?* (I have referred to the slightly amended 1948 edition for reasons that will become apparent.) This book was influential in attracting many young physicists to biological problems in the mid-twentieth century (Perutz, 1987), and indeed Gould (1995) considered it amongst the 'most important books in 20th century biology', while Paul Davies (2005) has emphasized the irony that 'one of the most influential physics books of the twentieth century was actually about biology'. Although many of its more original ideas turned out to be wrong, it was influential in a way that defies simple summary (see discussions by Perutz, 1987; Gould, 1995; Sarkar, 2013).

In his book Schrödinger (1948, p 71) famously described the process by which an organism survives as continually drawing negative entropy from the environment. In fact, on thermodynamical considerations this is not strictly true, as pointed out by Franz Simon soon after the book was first published. In the 1948 edition of his book Schrödinger added a note to this effect, admitting that it might be better to consider organisms as drawing on free energy rather than negative entropy. Free energy is the term used for energy that can be extracted from an environment by a particular process and is therefore available for use (Thorne and Blandford, 2017). These niceties of thermodynamic theory are not crucial for the concerns of this book. However, what Schrödinger was describing in thermodynamic terms is an important concept in ecology; namely that to survive, all organisms must acquire energy from their environment and, as no process can be 100% efficient, in so doing all organisms produce waste products that they release back into their surroundings. Indeed this is so fundamental that it could be considered *the* basic concept of ecology.

To a (very) naive physicist, organisms initially appear to break the second law; however, they are not closed systems as they draw free energy from their surroundings, and this is what enables them to give the *illusion* of cheating the second law. Just as a domestic refrigerator in keeping itself cool appears to violate the second law by moving heat from a cold to a hotter body, if the whole system is considered (in this example the refrigerator and its environment, which includes processes in the power station supplying the electricity) there is still a global increase in entropy from the operation of a refrigerator, although entropy decreases locally within the refrigerator. In a thermodynamic context the whole Earth system can be viewed in a similar way to an organism. For example Lenton (2004) described Gaia as 'a type of planetary-scale, open thermodynamic system, with abundant life supported by a flux of free energy from a nearby star'. In this respect it is sensible to consider Gaia as a superorganism, and the analogy is useful if one is considering free energy (or entropy). Indeed it is researchers who approach the Earth system through thermodynamics who currently seem most likely to take seriously the idea that there is something organism-like about Earth (e.g. Rubin et al., 2020; Rubin and Crucifix, 2022). For many biologists an immediate problem is that Gaia does not have obvious 'genetics', so while a classical biochemist (interested in metabolism, and thus effectively focusing on free energy use) could find merit in viewing the whole Earth system as a giant organism, a molecular biologist (interested in questions of information and genetics) would not. However, there is the possibility of the transfer of life from Earth to other planets, which although unlikely (at least outside our own solar system—see chapter 5) would potentially allow one to write about Gaia 'reproducing' and potentially having something analogous to genetics (Cazzolla Gatti, 2018). These questions about the suitability of identifying the whole Earth as an organism are very similar to debates from the early history of plant ecology where plant communities were sometimes described as like organisms. Arthur Tansley

(1920, 1935), while not viewing plant communities as organisms, conceded that they had some organism-like qualities and suggested that the term 'quasi-organism' might be appropriate—see van der Valk (2014) for a description of these historical debates. Indeed many ecosystem ecologists are now quite happy to write about 'ecosystem metabolism' and characterize ecosystems as 'autotrophic' or 'heterotrophic', as if they were organisms (e.g. Battin et al., 2023). Quasi-organism seems a good description of this approach to ecosystem ecology.

Like most metaphors, the idea that 'Earth is alive' is useful in thinking about some questions but misleading for others. It is no doubt relevant that James Lovelock's background was as a chemist who worked for many years in medical research, prior to the great expansion of molecular biology; indeed he was happy to describe himself as 'an old fashioned organic chemist' (Lovelock, pers. comm.). However, many of the most important applied problems in the environmental sciences are essentially ones related to Earth's metabolism: for example understanding changes in the carbon or nitrogen cycle. For these problems the metaphor of the living Earth has some merit, and in this context Tansley's term quasi-organism may be appropriate for the Earth system.

Schrödinger's point about organisms and thermodynamics can be simply characterized as:

$$\text{Energy} \rightarrow \text{organism} \rightarrow \text{waste product}$$

This is clearly very similar to the food chains common in introductory ecology texts. For example:

$$\text{Solar energy} \rightarrow \text{green plant} \rightarrow \text{herbivore} \rightarrow \text{carnivore}$$

However, in classical descriptions of food chains and food webs many of the waste products (CO_2, O_2, heat, etc.) are never shown, and many others (e.g. faeces or dead leaves) are only shown in special cases (Box 2.2).

Box 2.2 Illustrations of the partial nature of most food webs.

Here I briefly describe three food webs from different environments (terrestrial, freshwater, and estuarine) to illustrate the poor coverage of decomposers such as bacteria, protozoa, and fungi in most published food webs. Waste products are often ignored yet provide the free energy for crucial guilds of organisms (e.g. microbes involved in decomposition). All three of these examples were used in the first edition of this book; as Pringle (2020) has pointed out, over the last few decades we have added surprisingly few new data for well-studied food webs.

A terrestrial food web: Spitsbergen in the Arctic (Hodkinson et al., 2004)

This food web concentrates on the multicellular invertebrates. The main decomposers are treated as a black box labelled 'soil microflora'. The exception to this is a group of nine taxa of testate amoebae identified by myself from soil samples collected by Hodkinson and Coulson. However, these data are the product of examining only seven soil samples (not 42 samples as incorrectly implied by Hodkinson et al., 2004), while data on the macroscopic animals are based on substantially more work. The key point for the arguments in this book is that even in this highly detailed Arctic food web the crucial microorganisms receive little detailed attention.

A freshwater food web: Tuesday Lake, Michigan, USA (Cohen et al., 2003)

This study focused on the lake's pelagic food web. It excluded the, probably non-trivial, nutrient flow into the pelagic web from the littoral zone and the terrestrial system surrounding the lake. More directly relevant to the topic of this box, there are no microbial data (not even as a 'black box' as in the Spitsbergen example).

An estuary food web: Ythan Estuary, Scotland (Hall and Raffaelli, 1991)

This is an unusually complete food web; for example it contains 5,518 separate food chains. The web contains 92 species; however, microbes involved in decomposition within the system are relegated to a single black box labelled 'detritus'. Again this emphasizes the lack of detailed treatment of microbes even in unusually detailed food webs.

As it also follows from thermodynamic theory that no real system can be 100% efficient, any organism using free energy from its environment must be producing waste, which is then released back into the environment. As such there is nothing special in the fact that humans cause pollution (i.e. add waste to their environment); what is different about us is the size of the effect. Within my lifetime (I was born in 1963) the human population has more than doubled, and in addition an increasingly affluent and industrialized world is producing ever more waste. For example in 1990 there were approximately 478 million cars in the world; by 2000 this had risen to around 560 million; and Myers and Kent (2003) predicted that emerging major economies such as China would greatly increase this number over the next few years. This has indeed happened. One of my Chinese collaborators tells me that growing up in Eastern China he remembers mainly bicycles and motorbikes, with a few cars, during the 1990s but that over the last 20 years the number of cars has increased greatly (Jun Yang, pers. comm.). In fact, China has been the largest vehicle market in the world since 2009, with vehicle emissions becoming a major source of air pollution (Wang et al., 2019). Many other aspects of the Chinese economy have grown significantly over this period too, with associated environmental impacts (Figure 2.1). However, it's a complicated story; for example in recent years many goods produced in China have been for the American market, so this could be seen as effectively American-driven pollution (Gardner, 2018). In the vast majority of cases of human pollution the problem is one of rate or magnitude, as most of our waste products also occurred on Earth prior to our evolution. There are some rare exceptions to this, such as chlorofluorocarbons (CFCs), which have no natural sources (Lovelock et al., 1973), but normally with pollution we are altering the rate of production of a 'natural' product. However, rates and magnitudes matter—for example the caffeine in my coffee as I write this is

Figure 2.1 Rapid development in China over the last few decades has had significant environmental impacts. The photograph shows part of the building boom along Houxi River in 2015 in Jimei District, Xiamen City, southern China. The population of this district in 2000 was 148,000, by 2010 it was 580,000, and by 2020 it reached 1,037,000 (Mo et al., 2021). The environmental effects of this expansion in China are particularly obvious because of the rapid pace of change; in Europe and North America the effects tended to be spread across a longer time interval.

a pleasant (natural) stimulant, but at high enough concentrations it could kill me.

2.3 Sources of free energy

The source of energy we are most familiar with in Earth-based ecology is photosynthesis, which is considered in chapter 8. The other obvious source of energy for life as we know it on Earth are organic compounds, which would have been present on the early Earth and may have been the first energy source for life on our planet. Organic compounds would have existed on Earth before life; indeed these chemicals are widely distributed in space—for example a rain of interplanetary dust particles currently add approximately 300 tonnes of organic matter to Earth every year (Gribbin, 2000). Studies of Comet Halley on its last close approach to Earth in 1986 inferred the identity of a wide range of organic chemicals, from simple compounds such as pentyne and hexyne to more complex ones such as iminopropene and xanthine (Kissel and Krueger, 1987). More exotic energy sources than sunlight are also possible; for example it has been suggested that an ecosystem on Europa (one of the moons of Jupiter) would be theoretically possible even if the only liquid water is in complete darkness below the planet's surface. Rather than exploiting sunlight, such an ecosystem could be based on organic molecules that could be produced by charged particles accelerated in Jupiter's magnetosphere (Chyba, 2000; Chyba and Phillips, 2001).

2.4 Maximum entropy production and planetary ecology

Entropy may also be important in the context of this book as a way of understanding a planet's climate—clearly an important topic for ecology considered at a global scale. Classical thermodynamics tended to concentrate on equilibrium conditions, in much the same way that theoretical ecology focused on stable equilibriums in the mid part of the twentieth century. However, planets, like many ecological systems, are often non-equilibrium systems; this is especially true of planets likely to be of interest to a biologist (a 'heat death' planet is in a stable equilibrium condition, which makes for simplified mathematics but little chance of interesting ecology). During the 1970s it was realized that heat transport on some planets may be organized to produce entropy at the maximum possible rate (Paltridge, 1975). This observation, which seemed to fit data for Earth, has been controversial—not least because the mechanism has been far from clear and until recently these ideas have not been widely discussed outside a subset of climatologists and planetary scientists. As Lorenz (2002) has pointed out, 'much of the literature discussing the principle is mired in forbidding algebra', which has reduced its wider impact. Having struggled with the primary literature on this topic as well as attending several related workshops over the years, I can vouch for the truth of Lorenz's observation! However, some 20 years or so ago there started to be progress in producing a theoretical background for these observations that appeared to put them on a more accessible foundation (see Lorenz 2003 for a brief non-technical review and Dewar 2010 for more detail and a somewhat different approach from that reviewed by Lorenz). As Kleidon (2016) has pointed out, the renewed interest in these approaches since the early 2000s is mainly due to successful applications to Earth and some other planetary systems, combined with the theoretical advance cited above.

The simple box model of a planet's latitudinal heat flow in Figure 2.2 can be used to illustrate the main idea. In such a system the equation

$$dS/dt = F/T_1 - F/T_0$$

is maximized (where S is entropy and t is time; see legend of Figure 2.2 for remaining notation). If the heat flow (F) was zero then the entropy production would be zero. If the heat flow was maximized the whole planet would be at the same temperature ($T_0 = T_1$) and again production of entropy would be zero. However, between these two extremes, entropy production is positive and there is a single maximum value. It appears that planets (e.g. Earth, Mars, and Saturn's moon Titan) operate in such a maximum entropy state (Lorenz et al., 2001). One way of viewing such results is to suggest that the universal requirement for increased entropy (from the second law of thermodynamics) is responsible for the general circulation of the atmosphere on

Earth and other similar planets (Ozawa and Ohumura, 1997).

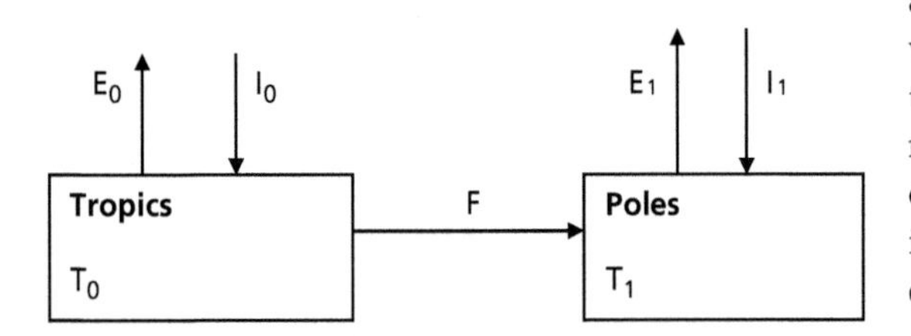

Figure 2.2 Two-box model of latitudinal heat flow on a planet such as Earth (after Lorenz, 2002 and Lorenz et al., 2001). The two boxes are of equal surface area; **Tropics** represents the planet's equatorial regions while **Poles** represents the sum of the two polar regions. Each box has an associated absorbed solar flux (I_0 or I_1) and also an outgoing thermal flux (E_0 or E_1). The latitudinal heat transport from tropics to poles is 'F'. If neither tropics nor poles are showing a net change in temperature (T) then:

$$I_0 - E_0 - F = 0$$

and

$$I_1 - E_1 + F = 0$$

In the context of this book the idea of maximum entropy production (MEP) raises a number of interesting questions. On Earth life is heavily involved in the dissipation of heat—for example there is the role of vegetation in affecting the land's albedo and the role of life in the regulation of greenhouse gases (Kleidon, 2016). If the MEP approach develops into a key way of thinking about planetary climates, then the presence of life on a planet will be a crucial part of the system. In addition the MEP approach is in some ways similar to the Gaia hypothesis, at least in as much as suggesting that a planet's system may be organized *as if* it was trying to maintain certain types of conditions over geological time. As such both ideas have attracted criticism because on a superficial reading they appear teleological. There were interesting initial attempts by Kleidon (2004) to integrate MEP and Gaia; based on thermodynamical considerations, he argued that the presence of life introduces additional degrees of freedom into the system and, as such, biotic activity should evolve towards MEP states.

2.5 Overview

Organisms maintain their highly ordered state by taking in free energy from their environment and (as use of this energy cannot be 100% efficient) releasing waste products back into the environment. This is so basic that it can be considered *the central concept of ecology*. The source of an organism's energy is a traditional topic of academic ecology; however, the nature and fate of the waste products is also of great importance and this has historically featured less prominently in ecological studies. The role of waste products, and their potential Gaian effect, will be considered in chapter 3. It is also possible that at a global scale ideas of MEP may be a useful approach in understanding a planet's climate; because of its role in heat dissipation, life can have a major effect in such systems. Most of these MEP-based ideas are very new and it is unclear how important they will ultimately be in understanding planetary scale processes. However, Chapman et al. (2016) showed that there had been a substantial increase in the application of MEP to ecology over the first 15 years of this century, which suggests it's an area that is growing in influence.

CHAPTER 3

Multiple guilds

3.1 The importance of waste

In chapter 2 I argued that the key idea at the heart of ecology is that organisms must use their environment both as a source of energy and as a dumping ground for their waste products. I described how these ecological ideas are linked to crucial concepts from the physical sciences such as the second law of thermodynamics and entropy. These arguments can be summarized as:

ENERGY→ **Organism** Some energy used →energy (waste)

As such there is still potentially free energy available in the waste products of organisms, as illustrated by the smaller font size used for the energy in waste products in this schematic representation. These waste products can potentially have important effects on other species, as illustrated by the problems caused by 'pollution', i.e. human waste products.

One approach to trying to visualize the quantity and importance of waste products produced by Earth's biota is to think about the temperate deciduous forests that dominated much of Europe, eastern North America, China, and Japan, before being reduced by human actions. Every year, at the end of summer, the leaves on the trees are transformed from the greens of chlorophyll to the reds and yellows of anthocyanins and carotenoids, before falling to the forest floor. These leaves, along with the other plant and animal debris produced over the year, amount to around 5 tonnes of biological waste per hectare per year. The figure for highly productive tropical forests can be an order of magnitude higher, at around 60 tonnes per hectare (Spooner and Roberts, 2005). It is an obvious observation that something must happen to get rid of this waste, since when we enter a forest we are not up to our necks in leaf litter!

The main strategy of this book is to use astrobiological thought experiments to give a wider perspective to thinking about Earth-bound ecology. As such it is worth considering the simplest possible biosphere, a hypothetical planet inhabited by only a single species; over time this species would convert its resources into both copies of itself and waste products. As individuals died, their remains would also be added to the waste product pool. Eventually such an ecological system would run down as the species ran out of resources, which had become locked up in waste products. As Tyler Volk (1998, p 52) has written, 'A monoculture planet is therefore a thought experiment with no place in reality'. I will return to thinking about this hypothetical planet in more detail in chapter 10.

Clearly in any sustainable ecological system it is necessary for material in waste products to be recycled to make them available for reuse (Figure 3.1). Any ecological system that produced an intractable waste product that couldn't be broken down would, on 'geological' timescales, be in trouble.

A useful concept here is the idea of an ecological guild—a group of organisms that all make their living in the same way. A classic example of a biochemical product that is unusually difficult to break down is lignin (Box 3.1). The key members of the modern lignin-degrading guild are fungi, particularly Basidiomyetes, many of which produce the familiar 'mushroom' fruit bodies. These 'white rot' fungi break down both lignin and cellulose (Figure 3.2). Their food source is the cellulose; however, they need to break down the lignin enzymatically to get at the cellulose. This process

The Fundamental Processes in Ecology. Second Edition. David M. Wilkinson, Oxford University Press.
 DOI: 10.1093/oso/9780192884640.003.0003

Figure 3.1 This beach on the Spanish island of Mallorca (photographed in 1999) shows what happens when a substance that is hard for organisms to break down (plastic) is added to a system over time. For several decades it has been possible to find plastic waste in quantity even on very remote beaches around the world (Barnes, 2002), as it is constantly being added to the marine system but only very slowly broken down. For example the naturalist and film-maker David Attenborough remembers seeing almost no plastic on the beaches of Aldabra in the Indian Ocean when filming there in 1983, but more recent visitors report 'humanity's rubbish on every part of the beaches' (Attenborough and Hughes, 2020).

leaves the remaining wood looking 'stringy, bleached and soft' (Spooner and Roberts, 2005). Had lignin-consuming organisms not evolved this could have proved problematic for abundant life on Earth, with increasing amounts of resources entombed in organic deposits and unavailable to life. While simple microbial systems may be able to continue in such conditions, it is difficult to see how an active and complex biosphere could be supported with such a lack of recycling.

Box 3.1 Lignin and why it's so hard to biodegrade

Lignins are phenylpropanoids and are the second most abundant polymer in most plants (after cellulose). They provide strengthening for cell walls and xylem vessels. They are also very important compounds in the Earth system; it is estimated that phenylpropanoids contain 30% of all organic carbon in the biosphere, most of this in the form of lignin (Davies, 2004).

There are three main reasons why lignin is difficult for organisms to break down (Robinson, 1990):

1) Molecules of lignin are insoluble in water and are too heterogeneous to be easily disassembled by specific enzymes.
2) Many lignin degradation products are themselves difficult to break down and/or are toxic to many organisms.
3) The C:N ratios in lignins are so high that few organisms could live on lignin even if they could break it down into usable compounds. Some of the fungi that can break down lignin get round this problem by trapping and digesting nematode worms using special hyphae,

so partly subsidizing their nitrogen intake. Examples of fungi that use this strategy include the oyster cap *Pleurotus* spp., which specialize in growing on wood (Spooner and Roberts, 2005).

Figure 3.2 Stinkhorn *Phallus impudicus* is a common species of 'white rot' fungus in Britain, mainly colonizing dead wood after it has fallen to the forest floor (Spooner and Roberts, 2005). Such lignin-degrading organisms are crucial in the breakdown of woody plant remains. In addition to its wood-rotting ability the fruit body of the fungus is famed for its phallic shape, celebrated in its scientific name, and its obnoxious smell which attracts flies and slugs to aid its spore dispersal (Ramsbottom, 1953).

3.2 The requirement for multiple guilds

Biological waste is also a potential resource for other organisms. For example, the carbon dioxide we breathe out is a raw material for photosynthesis by plants, just as the oxygen produced as a by-product of plants' photosynthesis is crucial to aerobic organisms like ourselves. The bodies of other organisms (alive or dead) are clearly composed of materials needed by life (that is why they are in an organism's body in the first place). In addition the fact that no energy extraction process is 100% efficient means that waste products of the organisms can provide a potential source of resources for other life forms, which will often quickly recycle these resources (Figure 3.3).

This last point is nicely illustrated by the effects of various veterinary antiparasitic drugs on the degradation of animal dung. One of the most well known of these drugs is ivermectin, which is used to control both endoparasites (e.g. lung worm) and ectoparasites (e.g. lice) in domestic animals such as cows, horses, and sheep (Floate et al., 2005; Sutton et al., 2014). Since the 1960s there has been a concern that dung containing these chemicals may break down more slowly, negatively affecting both plant growth on the affected pastures and the invertebrate species that specialize on feeding on animal dung (Table 3.1). In an experimental study in Britain, Sutton et al. (2014) found that ivermectin could still be detected in cow dung after 47 days (the maximum length of their experiments). While the effect has not been found in all studies, many researchers have found that dung containing chemicals such as ivermectin decomposes more slowly (Floate et al., 2005)—illustrating how one species' waste product can be a resource for another species. Lots of insects exploit mammalian dung, most famously the dung beetle. By dosing livestock with ivermectin to protect our livestock from parasitic insects, we have inadvertently made life harder for the dung beetles that are beneficial to the farmer by breaking the dung down to fertilize the soil, enhancing vegetation growth that the livestock then eats.

While it seems clear that these chemicals in dung can reduce the activity of invertebrates such as beetles and flies, surprisingly little is known about the potential effects on microbes and other organisms. Experiments by Sommer and Biddy (2002) showed that ivermectin still had an effect on dung degradation when larger invertebrates were excluded, but their experiment did not exclude organisms that could get into 1-mm-mesh bags. Therefore ivermectin could have been affecting very small animals and/or microbes. If, as most workers in the area seem to assume, the main effect of these chemicals on dung degradation is through their effects on invertebrates rather than microbes, then this system provides a good example of animals acting as 'catalysts' in decomposition, speeding up the process over the rate seen when just relying on microbes (see section 1.1). However, it is probably premature to rule out any effects of ivermectin on microbes without further study and the same is true of its possible effects on other organisms. For example, Porley and Hodgetts (2005) speculated that ivermectin may be involved in the decline of cruet collar-moss *Splachnum ampullaceum* in Britain. This moss grows on the fresh dung of grazing animals in wet habitats,

Figure 3.3 Egghead mottlegill *Panaeolus semiovatus* fruit bodies on cattle faeces in Cwm Idwal, North Wales, illustrating the important ecological idea that one organism's waste is another's energy source.

Table 3.1 Summary of total number of invertebrates recorded in control and ivermectin-contaminated experimental cow pats (dung) on a farm near Bristol in England. Numbers of coleoptera (beetles) are the sum of all individual larvae, pupae, and adults recorded, while figures for diptera (flies) are for larvae and pupae. The experiments were sampled over a time span of 20–100 days. Data are summarized from Wall and Strong (1987) who provide a more detailed tabulation.

Cow pat type	Number of samples	Coleoptera	Diptera
Control	16	1,229	286
Ivermectin	17	29	40

and as with the possible effects on microbes there is a lack of data to evaluate ivermectin's effects on mosses.

Whatever the role of microbes in the ivermectin example, a catalytic role for animals certainly appears to be the case with some more natural dung-based systems (Figure 3.4). For example, in eastern Kenya up to 48,000 beetles have been observed visiting a 3-kg pile of elephant *Loxodonta africana* dung during a two-hour period. During this time the entire pile of dung was eaten *in situ* or carried off to be buried by the beetles (Coe, 1987). This behaviour will greatly speed up the rate of microbial breakdown of the dung. However, it should be noted that much of the consumption of the dung by beetles will be facilitated by mutualistic symbiotic microorganisms living within the insects, so that a microbiologist would have some justification in viewing dung beetles as animated microbial systems!

These examples suggest that an ecological system needs organisms from at least two major guilds, autotrophs and decomposers (which release material from waste products, including dead autotrophs). Therefore any ecological system that has survived for a 'geological' period of time on a planet must consist of at least these two types of organisms. But what of other major guilds, such as predators and parasites?

3.3 Predators and parasites

Predators, and their prey, often feature very prominently in ecological studies. They are regularly the stars of television nature documentaries, which are (unfortunately) probably most people's main source of ecological knowledge. One of the most widely read serious 'popular' books on academic ecology over the last few decades has been the late Paul Colinvaux's *Why Big Fierce Animals Are Rare* (Colinvaux, 1980), which gives large predators pride of place in the title. Similarly, the most comprehensive single-volume textbook in ecology (Begon and Townsend, 2021) devotes 73 pages to predators but only 21 pages to decomposers. Yet predation, as a way of life, is not fundamental to ecology (unlike

Figure 3.4 African savanna elephant *Loxodonta africana* dung photographed in Kruger National Park, South Africa. In this case the obvious surviving vegetation fragments illustrate the idea that the utilization of food by elephants is a long way from 100% efficient.

decomposers). The grass–livestock–beetle system mentioned above would potentially function without any wolves (or humans). Indeed *if* there is life on other planets my expectation is that predation *sensu stricta* will only occur on a minority of them, since my guess is that most of them will not have progressed beyond bacteria-like organisms. This guess is based on the assumption that simpler types of life are likely to be more common than complex types, and the probability that eukaryotes took a considerable time to evolve on Earth. However, these assumptions could easily be undermined by a more detailed knowledge of the history of life on Earth or the discovery of life in other parts of the universe.

A reasonable definition of predation is that it is the 'action of one organism killing and eating another' (Krebs, 2009). In the history of life on Earth this probably required the evolution of phagocytosis, which is restricted to animals and protists (Margulis and Chapman, 2009). As such, predation *sensu stricta*, where the predator engulfs ('eats') the prey, would not have been present until evolution of the eukaryotes. Modern protozoa use phagocytosis to prey on a wide variety of organisms including bacteria, other protozoa, algae, fungal hyphae and spores, and small metazoans (e.g. see Gilbert et al., 2000 for testate amoebae and Foissner, 1998 for soil-living ciliates). The date of this major stage in evolution is not known with much certainty; however, currently the earliest convincing eukaryote fossils are 1.65 billion years old, while molecular clock approaches suggest a minimum age of 1.6 but a likely origin closer to 3.0 billion (Javaux, 2019; Cohen and Kodner, 2022).

One way to view phagocytosis is that it helps sidestep a potential problem with bacterial extracellular digestion. Prokaryotes are unable to take food particles directly into their cells, so they have to release enzymes into their environment to break their food down into chemicals that can enter their cells. Although there is a reasonably substantial literature on the structural changes that would have to evolve in order to allow phagocytosis, along with the phylogenetic implications (e.g. Keeling and Koonin, 2014), there is little written on the actual adaptive benefits. Two suggestions have been made in passing by theoreticians. One is the idea that the internal digestion allowed by phagocytosis prevents the loss of digestive enzymes to the organism's environment, and the potential for bacteria to 'cheat' by using enzymes produced by others, rather than synthesizing them themselves (Wilkinson, 2006; Sherratt et al., 2007). Alternatively, Tom Cavalier-Smith (2009, 2013) suggested that potentially the loss of the products of digestion—greatly reduced if digestion is internal to the cell—was the important driver for the evolution of phagocytosis. Consideration of the relative size of enzymes and digestion products suggests the mechanism proposed by Cavalier-Smith is likely more important that the enzyme-based ideas that I previously suggested. For small molecules, of the size often associated with nutrients and waste products, typical diffusion coefficients in water are around $10^{-9}m^2/s$, while proteins (such as enzymes) fall within values of $0.1–1^{-9}m^2/s$ (Dusenbery, 2009). So, loss of digestion products is likely the larger issue, although this is an area in need of formal modelling to investigate these ideas in more detail. However, it seems plausible to suggest that loss of the products of digestion, and/or the related enzymes, may have been one of the selection pressures leading to the evolution of phagocytosis. From an ecological perspective this produced the first true predation on Earth—with organisms capable of engulfing others.

Predators kill things in order to exploit them as a resource; parasites can exploit their prey without necessarily killing them (Box 3.2). While parasitism will harm their victim, and potentially may eventually cause the victim's death, that death is not essential to the parasite (but it is for the predator). Once autotrophs and decomposers exist then it is possible that parasites, capable of using resources from other living organisms while they are still alive, may have been quick to evolve. Observation of organisms on Earth shows that the lack of phagocytosis is no barrier to parasitism (e.g. parasitic viruses, bacteria, fungi, and plants; Figure 3.5). If Earth is at all typical then natural selection seems very powerful in targeting available resource pools, and so parasitism could be common to all biospheres. With Earth the only known example of a planet with life, how typical it is presents a big problem! However, selection is also powerful in targeting resources in computer-based artificial life models (e.g. Downing and Zvirinsky, 1999), suggesting that in this respect Earth may well be typical.

3.4 Parasites introduce a potentially important mechanism for density-dependent regulation

It is well known that parasites, especially some microparasites such as viruses and bacteria, can greatly affect the host's population size. For example the Black Death in the fourteenth century killed a substantial proportion of the human population of Europe; estimates of between 25 and 50% mortality are commonly cited (Cohen, 1995; Holmes, 2011). In some unfortunate places the mortality was much higher; one example was the Abbey of Meaux in northern England where only ten monks, out of an original complement of 42, survived the first wave of the Black Death (Schama, 2000). More recently a large number of people were killed in the influenza pandemic of 1918–1920. Historically estimates for the global death toll have been around 20 million (Anderson and May, 1991; Porter, 1997); however, more recent studies of contemporary records suggest a figure of 50 million or more (Johnson and Mueller, 2002). The ongoing COVID-19 pandemic (I am writing the first draft of this chapter in summer 2022) shows the power of infectious disease to still cause major human mortality and disruption. Crowded populations are more susceptible to such virulent microparasites; indeed this is why many authors assume that these microparasites were less important in human history before the development of agriculture and large settled populations (Anderson and May, 1991; Cohen, 1995; Diamond, 1997).

These examples of the effects of microparasites on populations of our own species illustrate the potential for parasites to reduce population sizes and potentially to act in a density-dependent manner—that is, to help regulate population size by increasing mortality at high population density while being a very minor threat at low population densities. Because of this it is very difficult for such a parasite to drive a host to extinction, at least in relatively simple mathematical models (Anderson and May, 1991). One potential exception to this simple theoretical story is parasites that show frequency-dependent transmission rather than density-dependent transmission. An obvious example are sexually transmitted diseases which can certainly drive hosts to extinction in mathematical models. Another exception is where the parasite has a wide range of hosts or even can survive by non-parasitic means, so density reduction of one host doesn't reduce the chances of infection. An example of this is the fungus *Batrachochytrium dendrobatidis* which has apparently caused extinctions of amphibian species in South America (where the fungus is introduced); in this case climate change

Box 3.2 Parasites and predators: problems of definition

Like most ecological definitions, 'parasite' and 'predator' are, at least in part, artificial classifications of convenience. While a protozoan engulfing a bacteria or a big cat eating a gazelle are clearly cases of predation, some examples are more difficult to categorize. To illustrate the problem two such cases are highlighted in this box.

Predation without engulfing the prey

Both carnivorous plants (Box Figure 3.2a) and nematode-trapping fungi catch and kill prey but extract the nutrients by extracellular digestion. As the whole organism is killed and consumed this would normally be described as predation rather than parasitism. Indeed Darwin (1875) wrote of insectivorous plants that they 'may be said to feed like an animal'. However, the vast majority of examples from natural history that are usually classed as predation involve the prey being engulfed.

Parasitism with (partial) engulfing of the prey

Many of the familiar multicellular parasites (especially ectoparasites such as fleas and lice) engulf a small part of their host without killing it. Such organisms are always traditionally classed as parasites. However, many herbivorous animals also only eat a small portion of the plant (without killing it). Arguably this is very similar to lice and fleas; however, caterpillars (Box Figure 3.2b) and deer are seldom described as parasites. One difference here is the nature of the prey. Plants are usually modular organisms, and loss of some modules (leaves or braches) will not kill them. This is much less the case with animals, although many invertebrates can lose legs, or some lizards their tails, without death ensuing.

Box Figure 3.2a A carnivorous plant, the pitcher plant *Sarracenia purpurea* growing on a 'floating bog' in Ontario, Canada

Box Figure 3.2b Parasite or predator? Caterpillars of the cinnabar moth *Tyria jacobaeae* feeding on ragwort *Jacobaea vulgaris* in Lincolnshire, eastern England

Figure 3.5 Toothwort *Lathraea squamaria* is a parasitic plant that does not photosynthesize but obtains its resources from the roots of other plants. This example was growing on Downe Bank (Figure 1.1) close to the former home of Charles Darwin.

appears to be exacerbating the problem (Pounds et al., 2006). In these cases there may be stronger selection pressures for the host to evolve tolerance to the pathogen (Searle and Christie, 2021). In a review of the empirical literature, de Castro and Bolker (2005) found very few well-documented cases of disease-induced extinction, suggesting it is probably rare in nature (exceptions may be when diseases affect populations that are already in trouble for other reasons).

The basic idea of density-dependent regulation has a history longer than that of academic ecology. For example it is implicit in the ideas of T.R. Malthus on human population at the end of the eighteenth century. His conclusions—about population crashes being likely at high numbers—are perhaps best known to biologists through their influence on the thinking of Charles Darwin (Mayr, 1982). One of the great advantages of mathematics over verbal reasoning is that it is usually much less ambiguous, such as the logistic equation—as featured in almost all ecology textbooks—which makes clear the assumption that population growth declines as population size increases (although it doesn't explain why). This equation (see glossary) was first used by Pierre-Francois Verhulst in the first part of the nineteenth century, although it was more famously rediscovered by Raymond Pearl and Lowell J. Read in the early twentieth century (Kingsland, 1995). Although parasites are not required for density-dependent regulation, they can certainly cause it, through affecting fecundity and/or survival. This regulation can be caused by macroparasites such as fleas, mites, and nematodes as well as the microparasites described above (Tompkins and Begon, 1999). Shortage of resources and/or increased 'pollution' from waste products when organisms are common can also have density-dependent effects and would even operate on a hypothetical planet populated by only one species, and so parasite free.

Density dependence is important in the context of this book's main thought experiment as it has the potential to have a positive Gaian effect. As such the presence of parasites on a planet may also tend to increase the probability of life surviving there for longer. This point is illustrated for single species populations by a simple Monte Carlo simulation model which shows the effects of adding density dependence on the probability of a population's survival. The model was run using VORTEX (Lacy, 2000; Miller and Lacy, 2003), one of the commonest population viability models used in conservation biology. The results (Figure 3.6) show that adding density dependence to the model increased the life expectancy of populations of a hypothetical species modelled to have a similar biology to the leopard (see legend of Figure 3.6 for more details).

A real-world example from conservation biology is an island population of a subspecies of silvereyes *Zosterops lateralis chlorocephalus* off the Australian coast. This is a small passerine bird species, and a population viability model of this island population was described by Brook and Kikkawa (1998), using an earlier version of VORTEX which did not have the facility to explicitly model density dependence. Their model suggested that this population was at some risk of extinction over a 100-year time period.

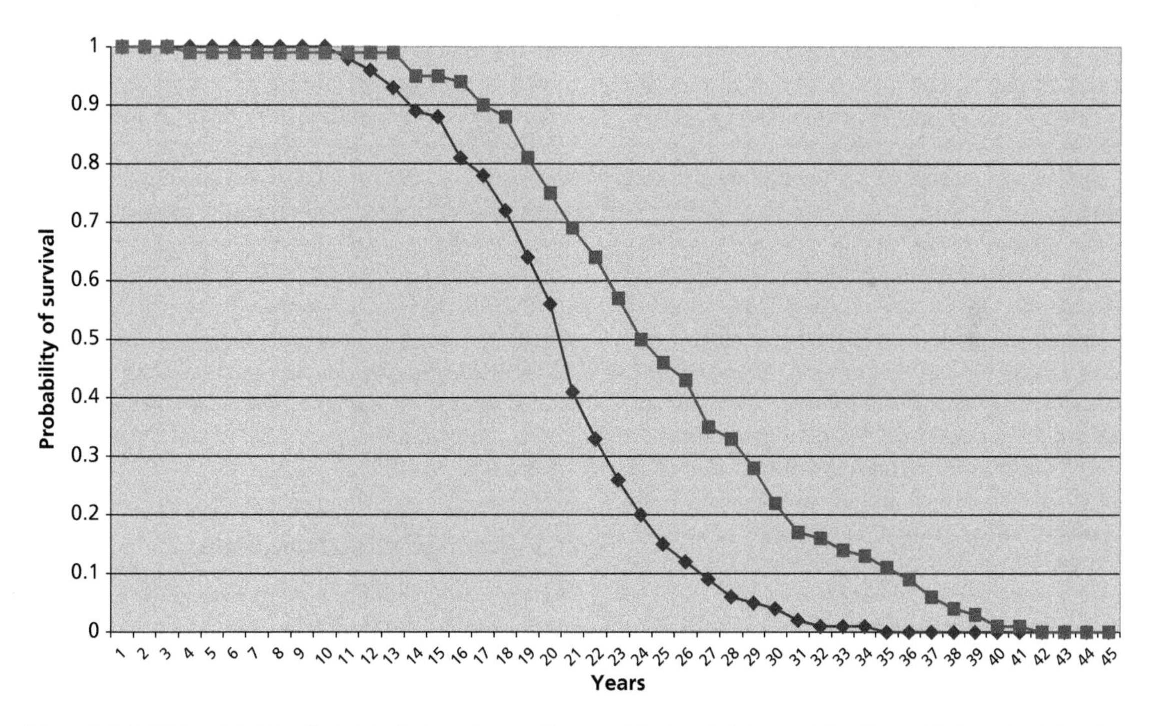

Figure 3.6 VORTEX model of the effect of density dependence on the probability of survival of a population. The modelled organism is based on the leopard *Panthera pardus*. Model variables follow Wilkinson and O'Regan (2003) except that a carrying capacity (K) of 200 was used (with a 50:50 sex ratio). In all runs of the model the population started out at K and the estimate of probability of the population's survival was based on 500 iterations. This model was run using VORTEX version 9.21 (Miller and Lacy, 2003). For the population marked by diamonds (♦) only a ceiling carrying capacity was used, while for the population marked by squares (■) full density dependence was modelled (but without an Allee effect). When an Allee effect was included (in a series of numerical experiments using a range of plausible values) the result was always intermediate between the results for ceiling carrying capacity and full density dependence without an Allee effect (to keep the graph simple these results are not shown). While there are many problems with using VORTEX to estimate the extinction probabilities of real populations (e.g. Harcourt, 1995; Coulson et al. 2001), here it is used in a qualitative manner to explore and illustrate the effects of altering model assumptions (see O'Regan et al., 2002 for an example of a study with a similar philosophy). The key point is that in this, admittedly simple, model the effect of adding density dependence is to increase the probability of population survival.

Subsequently McCallum et al. (2000) used an analysis of field data to demonstrate that this population exhibited density-dependent regulation, with a negative relationship between population size and population growth rate. They used their results to construct a simple density-dependent model of the population which predicted a much lower probability of extinction than did the original VORTEX model, which ignored density dependence. So, as in my simple model used to generate Figure 3.6, adding density dependence increased the predicted probability of survival of this population. Analogy suggests that density-dependent processes will also be important at a planetary scale, so potentially increasing the time span of life on a planet. One can imagine a microbial species becoming very common and potentially overexploiting its environment, only to be controlled by being targeted by a parasite. Many years ago J.B.S. Haldane (1932) argued that parasites played an important role in the evolution of species; they may be similarly important in the evolution of planets with life.

The discussion so far is an oversimplification, as density dependence does sometimes have the potential to *increase* the risk of population extinction—an obvious example being cases where increasing group size improves an individual's chance of survival, such as some group-living animals where, for example, a larger group may have greater chance of spotting predators (Allee et al., 1949). This process is usually referred to as the Allee effect, after Warder Clyde Allee who developed these ideas during the 1930s and 1940s. Since then many ecologists have assumed that this idea was of limited importance; however, during the 1990s there was a growing realization that the Allee effect may be important to the population dynamics of many organisms (Courchamp et al., 1999; Stephens and Sutherland, 1999; Bessa-Gomes et al., 2004). Much of this renewed interest has come from behavioural ecology; for example studies of African wild dogs *Lycaon pictus* suggest that because they are cooperative breeders they may exhibit a critical group size below which extinction becomes very likely (Courchamp et al., 1999). Unfortunately, due to our effects on the environment, many species of larger animals now live in small populations, and this is probably one of the main reasons for the recent growth in interest in the Allee effect.

Even if life is widespread on other planets, it is quite possible that species of the behavioural complexity of wild dogs may turn out to be very rare. However, the idea of the Allee effect also applies to organisms such as bacteria, which are more central to the key themes of this book. During the latter part of the twentieth century it became increasingly obvious that many microbes live in groups forming so called biofilms (Atlas and Bartha, 1998). Indeed current estimates are that 'up to 80%' of prokaryotic cells reside in biofilms (Penesyan et al., 2021). These microbes appear to benefit from growing in such aggradations, for example it can increase their resistance to antibiotics, both in medical situations and presumably also in the wild where other microbes may produce antibiotics as part of inter specific competition (Hoffman et al., 2005). As such these microbes may show inverse density dependence—the Allee effect—at small population sizes because of a reduced capability to resist antibiotics. Indeed the Allee effect is an area of growing interest in microbial ecology (e.g. Kaul et al., 2016). Some of the earliest uncontroversial evidence for life on Earth comes from the fossils of microbial colonies known as stromatolites (Figure 1.4), and it is easy to imagine that Allee effects could have been important in the survival of these colonies.

3.5 Other effects of parasites

As well as their effects on population dynamics parasites are important in an evolutionary context. Once parasites exist it is clearly in the interests of hosts to evolve strategies to protect themselves. For example the pupae of some species of ladybird—ladybug in North America—beetles (Coccinellidae) have evolved a defence against some species of parasitic flies and wasps which lay their eggs in the pupa (technically such insects are often called parasitoids). If the pupa detects the touch of such an insect, it can flick its anterior end quickly up and down, so making it more difficult for the parasite to lay its egg (Majerus, 2016). Ladybird pupae can sometimes be persuaded to exhibit this behaviour if experimentally touched with a blade of grass, mimicking the touch of a parasitoid.

More importantly, a good case has been made for parasites being involved in the evolution of sex, or at least the maintenance of sexual reproduction in populations once it has evolved. These ideas are particularly associated with W.D. Hamilton (see Hamilton, 2001 for collected relevant papers and Wilkinson, 2000 for a brief review of the history of these ideas). Hamilton argued that by repeatedly generating rare genotypes, sex made it harder for parasites to evolve adaptations to their hosts' defences—his argument was based on the essentially frequency-dependent ideas of J.B.S. Haldane who pointed out that parasites will mostly target common genotypes. These ideas have been very influential; however, a full explanation of the evolution and maintenance of sex probably requires additional ideas alongside Hamilton's parasite-based theory (West et al., 1999; Sherratt and Wilkinson, 2009), including ideas from genetics, which are not directly relevant to this book's ecological thought experiment. However, if Hamilton was at least partly correct then it raises the possibility that we would expect to see some form of sex on most planets with life. This is an idea that potentially ties into the earlier discussion of phagocytosis (see section 3.3). Eukaryotic sex requires cell fusion, which is not possible for prokaryotes with rigid cell walls. Cavalier-Smith (2009) speculated that 'Eating, not sex, generated the complexity of the eukaryotic cell', with cell fusion sex being an indirect consequence of the changes evolved to allow phagocytosis. If so, Cavalier-Smith's idea, with its emphasis on predation, is a very ecological explanation for the background conditions that allowed the origin of sex.

3.6 Overview

The problems faced by a hypothetical planet with only one species strongly suggest that any functioning ecological system must have organisms from at least two major ecological guilds, namely autotrophs and decomposers. While conventional predators do not seem to be crucial to planetary ecologies it is likely that parasites will quickly evolve and through density-dependent processes help to regulate population sizes. Density dependence—through lack of resources, build-up of waste products, or parasites—may be crucial in preventing the runaway population growth of a species, leading to it monopolizing a planet's ecology. While density-independent processes (be they a cold winter on a local scale or the impact of a large meteorite at the planetary scale) can greatly affect abundance they cannot provide regulation; this requires the 'thermostat'-like behaviour of density dependence. As such both multiple guilds and the presence of parasites are likely to have positive Gaian effects in most biospheres. Indeed parasites are likely to be an important selection pressure on most planets with life. This argument for multiple guilds has been based primarily on guilds defined by the way their members access carbon (as autotrophs or various guilds of consumers); however, similar arguments could be made for other types of biochemical guilds (*sensu* Volk, 1998)—for example the biochemical guilds of the nitrogen cycle such as nitrogen fixers and denitrifiers.

CHAPTER 4

Trade-offs and biodiversity

4.1 The problem of biodiversity

The diversity of nature, exemplified by Darwin's entangled bank, is so familiar from everyday experience that it is easy to accept it without much critical thought. We are so used to a multitude of species, and higher taxa, that it is difficult to appreciate the fundamental theoretical problems this diversity presents.

Consider these data from a 15 × 15 cm quadrat I recorded at an altitude of 1,890 m in the Dolomites (northern Italy). Within this very small area of ground in a coniferous forest I found mosses, sedges, grasses, and broadleaved plants from three different families (Asteraceae, Ranunculaceae, and Ericaceae) along with a small pine seedling. With the exception of the seedling, this quadrat was quite typical of the forest floor at this location. However, green plants can be considered as just ways for life to access solar energy, so why does it take so many different types of 'solar panels' to do this job on 225 cm^2 of ground in a forest on an Italian mountainside? An even smaller-scale example comes from the bark and wood chip mulch used on some of the soils on the science campus of my former University in Liverpool. Using a microscope I counted 100 individual testate amoebae (Figure 4.1) from a small sample of this mulch and found they belonged to at least 14 species from 10 different genera (since I made no attempt to split several difficult taxa which appear nearly identical under conventional light microscopy, the real species richness was probably slightly greater). Why is more than one species of protist needed to do the job of eating bacteria and other small organic particles in this apparently simple habitat?

These two examples illustrate biodiversity at the small scale—plants on 225 cm^2 of ground or protists in less than a gram of bark chippings. However, the problem is also apparent at the planetary scale. Robert May (1988), in a famous paper in *Science*, drew attention to the fact that we have very little idea how many species currently inhabit Earth. One result that most authors agree on is that insects appear to be especially rich in species; for example in the early 1980s Erwin (1982), in a well-known paper that's been cited over 1,500 times, estimated that there were at least 30 million insect species, a conclusion he reached by extrapolation from the ratio of described to undescribed species in a study of beetle species richness in trees in Panama. A decade later, using a similar extrapolation approach but based on data on hemipteran bugs in Indonesia, Hodkinson and Casson (1991) estimated a much lower figure of around 2 million. These estimates, along with a wide range of similar studies, raise two important points:

1) The range of estimated numbers of insect species strikingly illustrates our current lack of quantitative knowledge of biodiversity. Since the 1980s there has been a general consensus that the highest values are probably significant overestimates (Purvis and Hector, 2000; Novotny and Basset, 2005; Stork, 2018; Leather, 2022), although it's still impossible to answer the question 'How many insect species?' with any great confidence!
2) Even the lowest figures, such as Hodkinson and Casson's (1991) estimate of 1.8–2.6 million species, are still surprisingly large if one considers the obvious functional questions 'What do all these species do?' and 'Why are there so many?'

The Fundamental Processes in Ecology. Second Edition. David M. Wilkinson, Oxford University Press.
 DOI: 10.1093/oso/9780192884640.003.0004

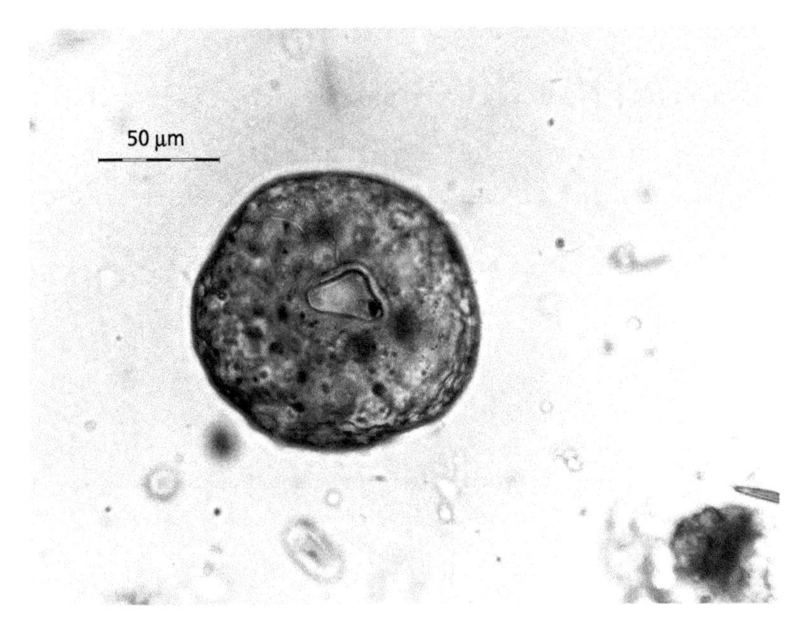

Figure 4.1 Shell of the testate amoeba *Trigonopyxis arcula*, one of the species found in the wood chip mulch as described in the main text (although this individual came from a moss sample from a Scottish mountain rather than urban Liverpool).

One of the reasons for concentrating on insects in the above analyses is that they form a significant proportion of all described species (Wilson, 1992); so if we had a good estimate for insects we could make an educated guess at the total species richness of all organisms on Earth. However, this assumes that the relatively low numbers of described species of microorganisms are a real measure of their diversity and not an artefact of them being small and difficult to study, a crucial point appreciated by May (1988) in his *Science* paper. The extent of geographical isolation, and hence presumably species richness, in microorganisms has been very controversial; this is not helped by the difficulty in defining 'a species' for many microbial groups. Consider the free-living protists, such as those described above from bark mulch. Some high-profile studies have claimed that almost all free-living protists show a complete lack of geographical isolation (see Finlay, 2002 and Finlay et al., 2004 for reviews); however, other authors suggested that while they were more likely to be cosmopolitan than larger organisms there was still evidence that many species have limited ranges (e.g. Foissner, 1999; Hillebrand et al., 2001; Wilkinson, 2001a). By analogy, what applies to free-living protists may also apply to other groups of microbes such as bacteria (Finlay and Clarke, 1999), so the outcome of this debate could greatly affect the estimated number of species on Earth, although there is always the possibility that different microbial groups may differ in their behaviour. As with the attempts to estimate insect species richness, over the last decade or two there has been some progress, and many people now seem to rule out the more extreme versions of extensive cosmopolitan distributions as the rule for microbes and other very small organisms (Fontaneto, 2011).

Until very recently a big problem has been that it was almost impossible to find very rare microorganisms—for example in a lake a microbe with an abundance of one individual per metre cubed would be almost impossible to find by microscopy, despite the fact that there might be a million or more individuals in the lake (Hutchinson, 1964). Molecular methods, especially high-throughput sequencing, now allow us to start to find this 'rare biosphere' (Pedrós-Alió, 2012). The fact that there appears to be geographical structure in the rare biosphere (e.g. Liu et al., 2015) tentatively suggests that even for these very rare taxa everything is not everywhere. That is, not all species are cosmopolitan. If free-living microbes are very species rich then calculations that assume insects make up approximately half of all species will be spectacularly wrong (this assumption would also be wrong if it turns out that the average insect has several species-specific parasites, although in this case insect species richness could still be used as a guide to total species richness). This debate also has implications for conservation biology; if some microbes have limited ranges then our actions could cause their extinction much more easily than if their ranges are global, with the species occurring anywhere in the world where the correct habitat is found (Finlay et al., 1997). For example, the testate amoeba *Nebela ansata* appears to be restricted to a small number of peatlands along the east coast of North America (Figure 4.2).

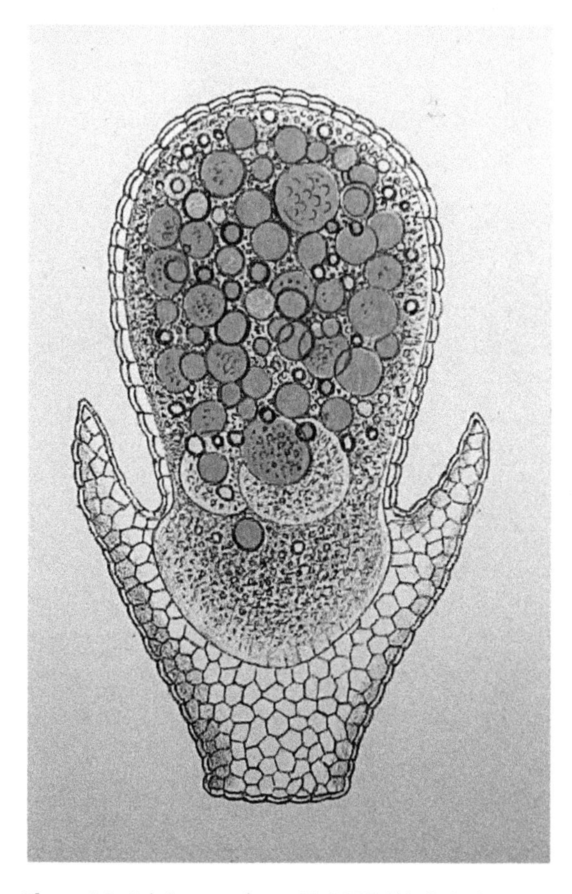

Figure 4.2 *Nebela ansata*, from Leidy (1879). This distinctive species of testate amoeba appears to be restricted to a limited number of peatlands along the east coast of North America. As some peatlands along the coast are thought to have survived through the last glacial maximum it's possible that this species survived the last glaciation in refugia along the coast and has since failed to recolonize much of North America, despite apparently extensive suitable habitat being available (Heger et al., 2011). Leidy collected this individual from bog moss *Sphagnum* in New Jersey, USA in 1877. He described it as 'a well fed individual', noting that all the food particles concealed the nucleus.

Chapter 3 discussed reasons why any planetary ecology should consist of two or more principal guilds, such as photosynthetic autotrophs or decomposers, and argued that multiple guilds are so fundamental that they would be found on any planet with life. However, those arguments give us no reason to expect that these guilds should be subdivided into a huge number of different types (species), which is what we observe on Earth. In this chapter, I consider one of the fundamental processes that can cause such biodiversity, namely the idea of trade-offs. Without trade-offs, the role of species-rich guilds could potentially be filled by single taxa, such as a 'perfect' photosynthetic plant (Figure 4.3). Such 'Darwinian demons' (Silvertown, 2005) would dominate a planet, leaving no room for speciation. I argue that these trade-offs would tend to cause within-guild biodiversity on any planet with life. I also consider the Gaian effect of this resulting biodiversity—a topic of more than academic interest, as we are currently in the process of greatly reducing biodiversity on Earth through our actions.

4.2 Trade-offs illustrated by human sporting performance

Consider the following description of a group of sportsmen: 'Of the eight, five are tall and slim, long-limbed and lightly-muscled; the other three are muscular and compact' (Perrin, 1990). This short 20-word description is enough to eliminate a number of possible sports; they clearly don't appear to be jockeys or sumo wrestlers, while the 'compact' individuals make it unlikely it's a description of a group of basketball players. That five are 'long-limbed and lightly-muscled' tends to rule out weightlifters, rugby players, or shot-put specialists. This brief fragment of prose is actually part of a description of a group of high-standard rock climbers at a cliff in the English Peak District in the 1980s and illustrates the point that different types of physique tend to be associated with success in different sports. To the uninitiated it may seem obvious that in rock climbing the more muscle you have the better, especially on overhanging rock. However, the main work done by a rock climber is to move their body weight up the rock face against gravity.

Figure 4.3 A snapshot of the diversity of flowering plants, illustrated by three native British species. Why are there so many species of plants? For example, there are around 1,560 native species in Britain (Stace and Crawley, 2015). Why not just have one generalist plant species and one species of pollinating insect? (a) Bee orchid *Ophrys apifera*; (b) bird's-eye primrose *Primula farinosa*; and (c) purple saxifrage *Saxifraga oppositifolia*.

Since muscle is a heavy tissue, what really matters is the climber's power to weight ratio; this is a trade-off between amount of muscle and body weight. So a good rock climber's body is more likely to look like that of a gymnast or professional dancer than a weightlifter or bodybuilder.

These ideas have been developed with more quantitative rigour in an analysis of the performance of decathletes by Raoul Van Damme and colleagues. The decathlon comprises ten different track and field events and Van Damme et al. (2002) constructed a data set of the performance of 600 'world-class decathletes' from information available on the Internet. The effects of trade-offs were apparent in the comparison of performance in some of the events. For example, there was a negative correlation between performance in the 100 m and the 1,500 m, also between performance in the shot-put and 1,500 m. Overall there was a highly significant negative correlation between overall performance across the ten events and excellence in a particular specialist discipline. The sorts of people who tend to win decathlons do well in all the disciplines but may lose to a specialist in any particular event. Van Damme et al. (2002) explicitly used their analysis as a novel way of quantifying the effects of specialization and trade-offs in organisms. They concluded that 'in an environment in which the selection criterion is combined high performance across multiple tasks, increased performance in one function may impede performance in others'.

4.3 Trade-offs in ecology

In both sport science and ecology a good definition of trade-off is 'any situation in which the quality of one thing must be decreased for another to increase'. Peter Grubb (2016), who recommends this definition for use in ecology, further draws a distinction between what he calls 'true trade-off', where as one feature (A) increases another feature (B) must necessarily decline, and 'boundary line' trade-off, where more variation is possible but the trade-off sets an upper boundary to possible relationships between A and B. In this context the sporting examples discussed in the previous section are effectively boundary line trade-offs as it's possible to find people excelling in a particular sport despite atypical body types—however, a jockey is never going to excel in sumo wrestling! However, the key point of trade-offs in general is that constraints matter (e.g. height in basketball players). The idea of trade-off appears to have grown in importance in ecological theory over the last couple of decades of the twentieth century, as illustrated by an increase in the use of the term in the ecological literature after 1990 (Figure 4.4). As Grubb pointed out, when Cherrett (1989) compiled his list of important concepts in ecology—as identified in a poll of ecologists—'trade-off' and 'constraint' didn't feature; Grubb (2016, p 25) commented, 'Surely they would both be in the top 50 now. The two concepts are central to current ways of thinking'.

Typical ecological examples are seed and spore size, where there can be a trade-off between size and numbers produced. For example 97 species of wood-inhabiting fungi in Finland showed a negative correlation between the amount of spores released and spore size (Norros et al., 2023). Another good ecological example of trade-offs in action is background matching in cryptic animals (Figure 4.5). Many species of moth have patterns that match part of their environment, such as the oak beauty *Biston strataria* which matches lichen-covered tree trunks (Cott, 1940), while other British moths, such as the early thorn *Selenia dentaria* and lappet *Gastropacha quercifolia*, look like dead leaves (Majerus, 2002). Such insects have a problem—their environment is not just made up of either lichen-covered trunks or nothing but dead leaves; instead it comprises many other potential backgrounds such as green leaves of various shades, lichen-free branches, and colourful flowers. Which background should they match?

With a simple mathematical model (Box 4.1) it is possible to investigate the conditions under which an insect should specialize in matching a particular background or have a compromise colouration where it is a less good (but not terrible) match to a wider range of backgrounds. There are trade-offs here; having a good match for one background can limit the range of microhabitats an animal can safely use. This is important as heterogeneous environments are common in nature, while environments of uniform colour and texture appear to be much rarer. There is evidence from experiments with captive great tits *Parus major* feeding on artificial prey that under some conditions compromise colouration, which is a reasonable match for more than one background, is a viable answer for cryptic prey living in heterogeneous environments (Merilaita et al., 2001), as predicted by the mathematical model outlined in Box 4.1. Using the decathlon analogy, one of the specialist moths will be harder to see than a moth with a more generalized pattern in the specialist's correct microhabitat (it will outcompete the generalist in that particular specialized 'event'), but the more generalist moths may outcompete the

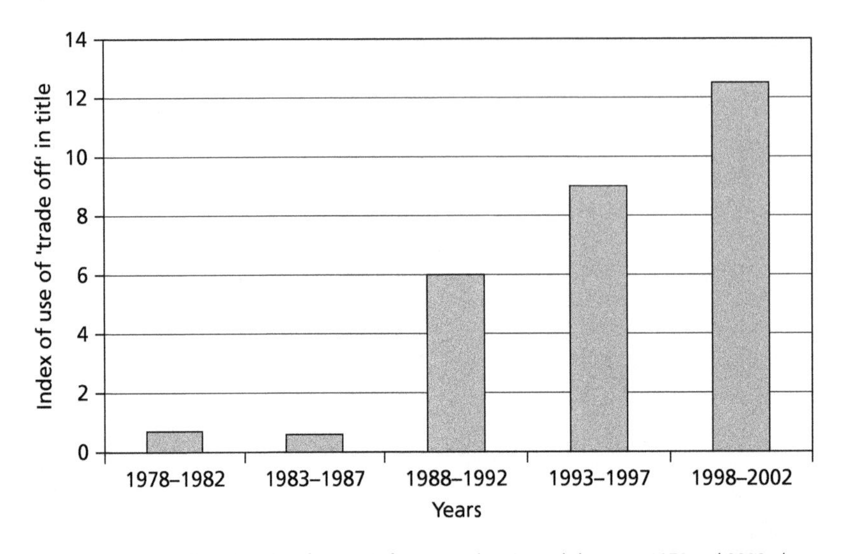

Figure 4.4 The use of the word 'trade off' in the titles of papers in five top ecology journals between 1978 and 2002; data are taken from appendix 1 of Nobis and Wohlgemuth (2004). The journals used were *Ecology*, *Journal of Animal Ecology*, *Journal of Ecology*, *Oecologia*, and *Oikos*. All showed an increase in number of papers published per year over this time period; to remove this effect I have divided the number of papers mentioning trade-off in each time period by the total number of papers published in that period, then multiplied by 1,000 to give an index based on whole numbers. Note the increase in explicit use of the word 'trade-off' since around 1990. I have not attempted to update these data for the second edition of this book because I suspect some paper titles that would have used the term 'trade-off' in the past instead now use the related (and currently fashionable) idea of 'traits' in their titles.

Figure 4.5 At its best background matching can be extraordinarily successful. There are two ptarmigan *Lagopus mutus* in the centre of this photograph of a Scottish mountainside; however, they are so well camouflaged that I considered using another photograph to illustrate this point where the organisms would be easier to see. Indeed the Japanese translator of the first edition told me that he only managed to spot the birds when he saw a colour version of the photograph (all photos in the first edition of this book were reproduced in black and white). This species of grouse increases its background matching by having a white plumage in winter and grey plumage—as in this photograph—in summer (Cramp and Simmons, 1979); however, against the 'wrong' background it will stand out (e.g. a grey summer bird against a white snow patch or a white winter bird against grey rock and lichen). As such there are potential costs to accurate background matching as well as benefits.

specialist if the 'game' is played over a wider range of microhabitats (so winning the gold medal for overall performance). However, the environmental conditions matter; in models of these systems using humans as the 'predators' in a computer game, colleagues and I found that specialist background-matching prey often did better than a compromise colouration (Sherratt et al., 2007).

Box 4.1 A simple model of crypsis in an environment with multiple backgrounds

This box is based on the generalized version of the model of Merilaita et al. (1999), as described by Ruxton et al. (2018). It provides a simple model to illustrate the idea of trade-offs and when an organism should evolve to specialize and when it should be a generalist.

Consider two backgrounds (a and b); these could be lichen-covered bark and lichen-free bark for the case of a moth that spends time on trees. Let the probability of a moth being viewed by a predator against background 'a' be V_a; similarly V_b is the probability of being seen against background 'b'. Note the predator may or may not actually detect the 'moth'. The corresponding probabilities of not being seen against these two backgrounds are C_a or C_b.

Since in this simple example the prey can only be on one of these two backgrounds:

$$V_a + V_b = 1 \quad \text{(remember as these are probabilities they must add up to 1)}$$

The overall probability of being detected by the predator (D) is:

$$D = V_a(1 - C_a) + V_b(1 - C_b)$$

As any moth must be either seen or not seen by the predator the probability of escaping detection (E) is:

$$E = 1 - D \qquad \text{(since } D + E = 1\text{)}$$

In this example I am interested in the trade-off between crypsis against two backgrounds and improved crypsis on one background. In this case, C_b is a declining function of C_a; that is as the chance of the moth not being seen against background a improves its chance of being seen against background b increases. The maths is slightly more complex than the simple statements of probabilities above, but the important result can be seen in Box Figure 4.1 (if you are prepared to take the maths 'on trust', ignore the following equations). Formally:

$$C_b = f(C_a),$$

$$df(C_a)/dC_a < 0 \quad \text{(this is just stating that as } C_a \text{ increases, } C_b \text{ decreases)}$$

With a bit more maths it can be shown that E has a maximum or minimum if:

$$df(Ca)/dCa = -V_a/V_b$$

This will be a maximum if:

$$d^2f(C_a)/dC^2_a < 0 \quad \text{(the curve this produces is shown on the graph)}$$

As can be seen from Box Figure 4.1 a series of outcomes are possible: either be a generalist and try to look a bit like both backgrounds or specialize in background a or b. Which outcome is best depends on the actual values for V and C in any given example.

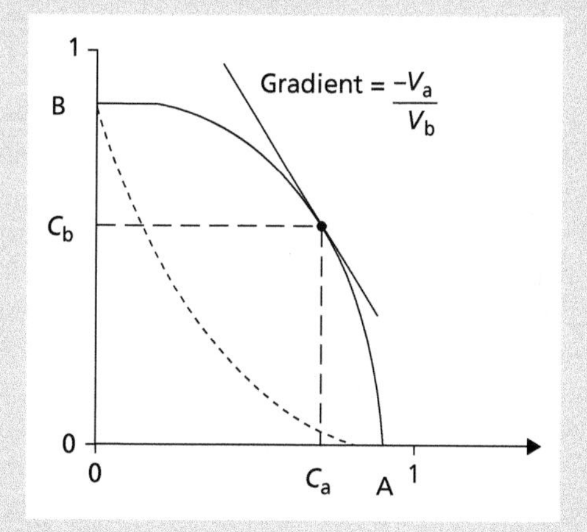

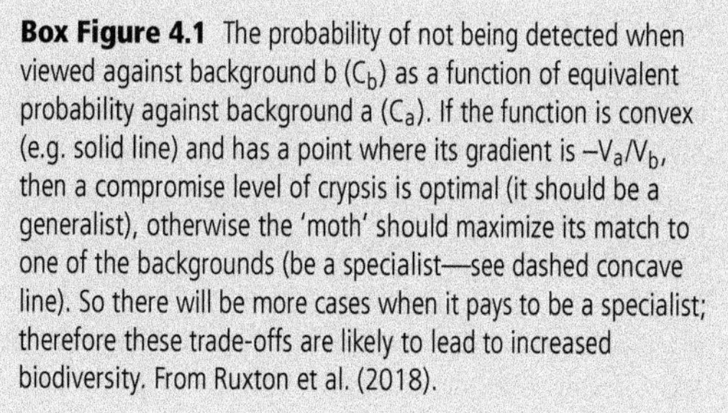

Box Figure 4.1 The probability of not being detected when viewed against background b (C_b) as a function of equivalent probability against background a (C_a). If the function is convex (e.g. solid line) and has a point where its gradient is $-V_a/V_b$, then a compromise level of crypsis is optimal (it should be a generalist), otherwise the 'moth' should maximize its match to one of the backgrounds (be a specialist—see dashed concave line). So there will be more cases when it pays to be a specialist; therefore these trade-offs are likely to lead to increased biodiversity. From Ruxton et al. (2018).

This idea of generalist and specialist species is nicely illustrated by some members of the Corvidae, the bird family containing crows, jays, and magpies. The raven *Corvus corax* is widely distributed across the Holarctic (Goodwin, 1986) and seems able to survive in a wide range of climates and habitats. I have watched ravens on cold wet hillsides in Britain and in the hot desert mountains surrounding Death Valley in California. They do well in wild country where you can walk for hours without seeing another human, but I have also watched them in the middle of alpine villages and on urban-fringe refuse dumps in Mediterranean Europe. It is interesting to contrast the raven with the nutcracker *Nucifraga caryocatactes*, a dark-brown jay-like bird with white spots; this is a specialist, largely restricted to a single habitat—montane coniferous forests across Eurasia (Coombs, 1978; Goodwin, 1986). Other specialist members of the Corvidae include the red-billed chough *Pyrrhocorax pyrrhocorax*, which uses its slender bill to dig invertebrates—often ants—from the soil, usually in well-grazed areas with close-cropped vegetation, and the closely related alpine chough *Pyrrhocorax graculus*, a montane specialist which has been recorded feeding on food dropped by mountaineers near the summit of Everest in the Himalayas (Coombs, 1978; Cramp and Perrins, 1994).

In the context of the thought experiment I have used to structure this book, an obvious question is 'What process leads to specialization by organisms (and hence to biodiversity) and would they be expected to operate in all conceivable ecologies?' Superficially it may seem that the best idea is to be a generalist. However, as the decathlon example illustrates, a generalist will usually be outcompeted by a specialist under specific conditions. Generalists win the decathlon because the rules require performance in ten *different* events. The raven illustrates this point about competition with specialists. In North America you often see ravens in desert habitats (Wilkinson, pers. obs.; MacMahon, 1985); however, it is not usually found in the deserts of North Africa or the Middle East, where it is replaced by related species from the same genus which appear to be more specialized for desert living (Goodwin, 1986). The athletic analogy is of a great decathlete losing a particular event, say the shot-put, to a lesser athlete who specializes in that particular event.

The decathlon example illustrates another important point: trade-offs limit the extent of the generalist strategy. While one person can be reasonably good at all the events in the decathlon, if the sport was broadened to include the skills of a jockey and a sumo wrestler then no one could do well in all of the events. Ravens are found in a wide range of habitats and climates but are absent from others, such as the humid tropics. Trade-offs make a superorganism or Darwinian demon, which is good at everything, a figment of the science-fiction writer's imagination.

4.4 Trade-offs and biodiversity

The idea that trade-offs contribute to the generation of biodiversity is not new (e.g. Tilman, 2000); however, it has been given relatively little prominence in academic ecology. For example, for the first edition of this book I searched for the term 'trade-off' in the indexes of the current editions of ten university-level ecology textbooks (published between 1993 and 2006) and found that only two texts indexed the term. Things have now improved somewhat, with trade-off receiving more attention in many current textbooks—for example Begon and Townsend (2021) indexes over 20 references to trade-offs across the book's 844 pages. The main aim of this chapter is to argue for the central importance of trade-offs in ecological theory; as such it should feature even more prominently in textbooks.

As argued above, at the most basic level trade-offs prevent an organism from being good at everything, and so help rule out persistent planetary ecologies based on a single species. Within the major guilds described in chapter 3, trade-offs also make it unlikely that a single species, such as a single taxon of photosynthetic autotroph, will dominate the system. Because of trade-offs a single photosynthetic taxon cannot perform equally well in a wide range of light levels or, if terrestrial, under very different levels of water availability. Even in my small patch in the Dolomites light levels and other factors will vary with the seasons. As such within any major guild I would expect trade-offs to lead to a proliferation of species on any planet with life.

Trade-offs preventing one species outcompeting all others are therefore a good candidate for the most fundamental reason for the development of biodiversity; however, they clearly don't explain all the diversity of life on Earth. For example, dispersal (see chapter 5)—or lack of it—and other aspects of biogeography play an important role in generating biodiversity on Earth. Some of biodiversity can be put down to different species filling the same role in different localities separated by some barrier than prevents one species spreading from one locality to another. This is illustrated by the distribution of several families of flightless birds; the rheas (Rheidae) are found in South America, while the ostrich (Strumionidae) is now found in Africa but was previously also in parts of Eurasia, and the emus (Dromaiidae) are restricted to Australia (del Hoyo et al., 1992). These distribution patterns (with the proliferation of families) are clearly more to do with accidents of evolutionary history than trade-offs (see chapter 11 for a more detailed discussion of historical contingence). However, the convergent characteristics of these three families—along with the cassowaries (Casuariidae), kiwis (Apterygidae), and several extinct taxa—are probably partly due to trade-off constraints as well as shared phylogenies.

While trade-offs provide an important theoretical reason for biodiversity, which should operate in all ecologies, the theory is currently qualitative. We have no quantitative theoretical understanding that would allow us to predict the species richness of a planet, even to an order of magnitude! Such predictions may be impossible if historical contingency is important on most planets; however, making these ideas more quantitative would surely be a useful step forward in our ecological understanding.

4.5 The Gaian effect of biodiversity

The ecological role of biodiversity, usually defined as species richness for convenience of measurement, became a highly controversial subject at the end of the twentieth century (Grime, 1997). This debate has a longer history (e.g. Elton, 1958); however, the emphasis changed from the mid to the late twentieth century. When Robert May (1973) wrote his, now classic, book on *Stability and Complexity in Model Ecosystems* he followed Elton in emphasizing questions in population ecology, such as 'Are populations in complex ecosystems more stable?' However, for mathematical convenience May considered small perturbations to systems that had come to a settled equilibrium, which are probably not representative of many systems in planetary ecology. More recently the trend has been to start to ask questions about the effect of species richness on ecosystem services (e.g. Naeem et al., 1994; Tilman et al., 1996). As is now well known, these questions are not straightforward to answer. For example, the more species there are in a system then the more likely it is that one of them will be highly productive and so give the impression that higher species richness causes higher biomass (an effect described as a 'hidden treatment' by Huston, 1997), so species richness cannot easily be separated from the identity of the species. These questions are now of great importance due to large-scale human-caused extinctions. So, what are the Gaian effects of biodiversity?

Since plants are basic to most terrestrial food webs and are easier to experimentally manipulate than most animals, they are a useful group to think about. Consider some measure of the 'performance' of a plant community, such as total biomass. What is the effect of the number of species on biomass? One possible answer comes from economics and is often known as the portfolio effect. The idea is that a cautious investor in a stock market is more likely to sleep soundly at night if they put their money into a portfolio of shares rather than investing in shares from a single company. Statistically a portfolio investment will show less variance—a lower chance of either very high profits or spectacular losses—than an investment in a single company's shares. This is due to statistical averaging and is effectively an economic application of the idea of trade-offs—investing in a portfolio trades off the chance of very high returns against a reduced probability of large losses.

The idea is illustrated by Figure 4.6, which shows the effect of investing £7,000 in shares of computing firms quoted on the London Stock Exchange in early 2005. If you had been lucky (or skilful?) and invested all your money in the best-performing company, then after a four-month period you would have made over £500 profit. However, if you had chosen the worst-performing company you would have made a loss of over £1,000! A portfolio investment where your £7,000 was spread equally between seven companies would have given you a much more modest loss.

The means and standard deviations of these share prices over 16 weeks are given in Table 4.1; in

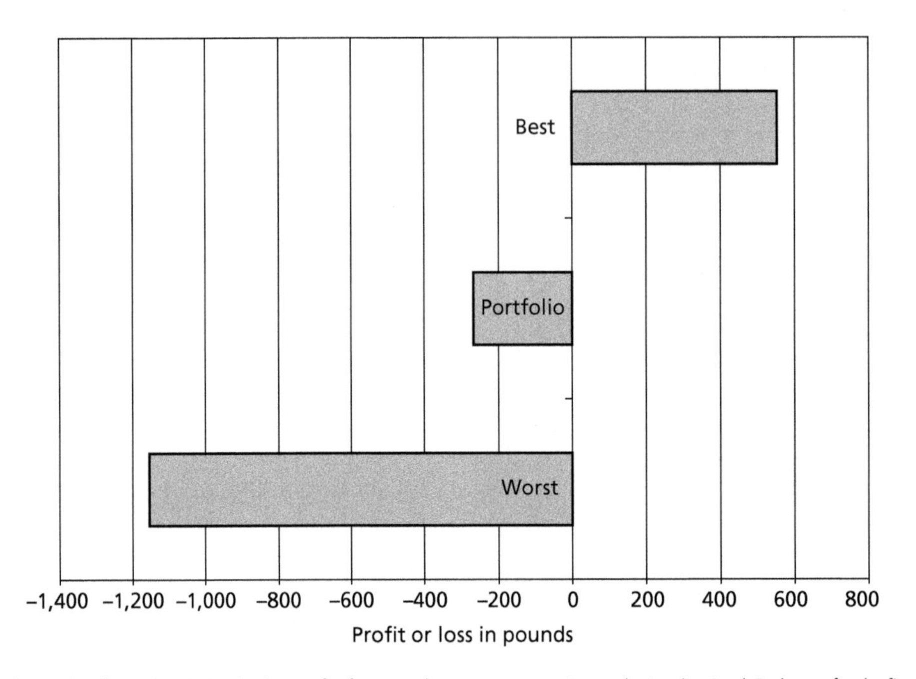

Figure 4.6 The results of investing £7,000 in shares of software and computer companies on the London Stock Exchange for the first 16 weeks of 2005. The portfolio is made up by investing £1,000 in each of seven firms (this made a modest loss). Also shown is the outcome of investing £7,000 in the best performing of these seven companies and in the worst performing. Mean and standard deviations of these share prices (sampled once a week during this period) are given in Table 4.1. Given the poor performance of most of these shares over this period I am pleased that I did this as a thought experiment rather than a real experiment!

this case the standard deviations of the worst-performing shares and the portfolio are similar (the worst shares were very steady for much of the time, then fell sharply). However, the best-performing company shows a much larger variation (if you had sold your shares in this firm at any point over the 16 weeks you would always have made a profit, but the size of that potential profit changed greatly from week to week). The portfolio of seven companies would have given you either a modest profit or a modest loss over the 16-week period depending on when you sold your shares. This extended example may seem overly detailed to illustrate such a simple idea; however, it is likely that this portfolio effect is an important ecological process.

In an ecological context the suggestion is that a species-rich community will also show less variation in measures of performance (such as biomass) for the same reasons (Doak et al., 1998). For example, Lázaro et al. (2022) recently showed that the portfolio effect had a role in stabilizing pollination services, that is maintaining a supply of pollinators, in a Mediterranean plant community. It's important to realize that this effect is a statistical relationship; put more colloquially, on average the community will be more stable, but there will be exceptions that prove the rule. This is shown by the economic data in Table 4.1 where the lowest standard deviation was actually for the worst-performing shares, not the portfolio, for reasons described above. This idea of statistical averaging is most easily understood by considering a monoculture; in such a situation if the single plant species

Table 4.1 Mean and standard deviations of share prices (in £) measured once a week over a 16-week period, illustrating the portfolio effect. See caption of Figure 4.3 for further details.

	Mean	Standard deviation
Worst shares	35.9	3.48
Best shares	361	9.06
Portfolio	194.9	3.92

is not growing well then the total biomass of the system is likely to decline—the same would be true of other systems such as a microbial mat (potentially of relevance to a wider range of planets than a 'plant' community). If, however, this species is part of a more diverse community and it performs poorly then other species may perform better and, at least in part, take its place; so there may be no decline in biomass. It seems very likely that this statistical averaging (the portfolio effect) will operate in many ecological communities and indeed also at the planetary scale, where total biomass can be very important in the 'physiology' of a planet (see chapter 7). The extent of this importance may vary between different systems and could depend on the exact nature of the relationship between variance in species abundance and mean species abundance (Tilman et al., 1998). In the simplest case, if the performance of every plant (or microbe or other organism) was positively correlated then the idea would not work, just as a portfolio of shares may not protect you from financial loss during a stock market crash when almost all shares are declining in value. A nice experimental illustration of the portfolio effect was provided by Dang et al. (2005); they studied fungal decomposers in stream microcosms and found that reducing fungal diversity led to increased variability in rate of decomposition, as would be predicted by a statistical averaging approach.

There are many other reasons why species richness may contribute to the long-term persistence of an ecological system. Once again consider a plant community (or a biofilm if you are a microbiologist); anyone who has spent time describing plant community data—such as the quadrat from the Dolomites described in section 4.1—will be familiar with 'lost plants' (*sensu* Janzen, 1986). These are plants that exist in low numbers in a plant community, often as seedlings or stunted individuals of species that are either absent or very rare as fully grown individuals in the community. The pine seedling in the Dolomite quadrat is an example of this as the main tree in this part of the forest was spruce *Picea abies*; the pines dominated the forest higher up the mountainside around the tree line. Janzen called these 'lost plants', by analogy with vagrant animals such as the yellow-bellied sapsucker *Sphyrapicus varius* (a North American woodpecker) and monarch butterfly *Danaus plexippus* (again from the Americas); both of these species occasionally turn up in Britain after having accidentally crossed the Atlantic.

Such rare organisms are often considered to be of no ecological importance. However, at least in the case of plants, Philip Grime (1998, 2001) has argued that they could be crucial in allowing communities to cope with climate and other changes. Grime called these species 'transients' and suggested that a species-rich community with more transients was more likely to contain species with the correct attributes to suit new conditions. As such, species-rich communities may be more able to cope with climate change. In this case species richness increases the probability of a species with particular characteristics being present in the system. This is a real ecological effect, not something that should be written off as an accidental effect—or 'hidden treatment' in an experimental context (c.f. Huston, 1997). A very similar idea, predating Grime, was suggested by James Lovelock (1992); in discussing the role of biodiversity in Daisyworld models (described in chapter 7) he wrote, 'destroying biodiversity will reduce the reservoir of apparently redundant, or rare species. Amongst these may be those able to flourish and sustain the ecosystem when the next perturbation occurs'. The tendency to dismiss species as redundant, as far as ecosystem function is concerned, may be partly a product of thinking about them at the wrong timescale. On the three- or four-year timescale of many PhD studies these species may be doing nothing of importance, but on a longer timescale with climate and other changes this may not be the case. This is one of the reasons that long-term ecological studies are so important, albeit difficult to fund (Wilkinson, 2021b).

A problem with these ideas is that they make predictions about the behaviour of communities over longer timescales (perhaps tens to thousands of years), which makes them hard to study. An obvious approach is to reconstruct past changes using palaeoecological methods. The relevance of this approach to 'lost plants' is illustrated by data from a study of a peatland at Astley Moss in northwest England by Steve Davis and myself (Davis and

Wilkinson, 2004). We reconstructed the bog vegetation over the last 5,000 years by looking at the remains of plant leaves and roots preserved in the peat. Over time the vegetation alternated between communities dominated by bog moss *Sphagnum* spp. and others dominated by heathers (*Calluna vulgaris* and other species). When *Sphagnum* dominated the community, heathers formed a minor component of the vegetation and vice versa (Figure 4.7), so as conditions changed species that had been rare in the vegetation took over as dominants, maintaining vegetation cover at the site. Similar patterns could also be seen in the abundance of testate amoebae whose shells are preserved in peats along with the plant remains; over time different species were dominant, with pronounced changes in the community around 2,500 years ago, probably due to climate change. Clearly the analysis of palaeoecological data has promise in investigating ideas about the long-term role of rare species in communities.

In an illuminating discussion of ecosystems from the perspective of the engineer's concept of redundancy, Shahid Naeem (1998) pointed out that a basic principle of reliability engineering is that 'the probability of reliable systems performance is closely tied to the redundancy in its design'.

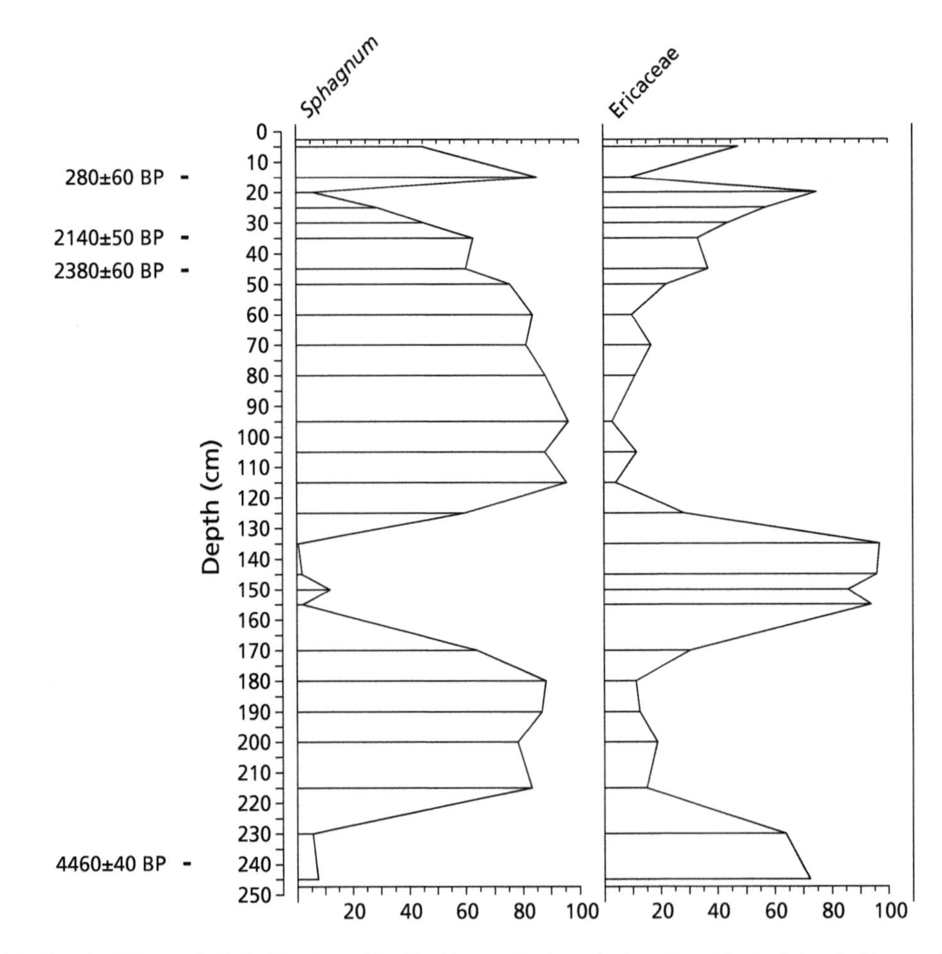

Figure 4.7 Macrofossil diagram for Astley Moss (part of the Chat Moss peatland complex in northwest England, described in more detail in section 9.1), showing the abundance of *Sphagnum* moss and Ericaceae (the heather family) remains in the peat. This record covers approximately 5,000 years; dates down the left side of the diagram are in 'radiocarbon' years before present (BP—where present is defined as 1950). Calendar years tend to be a little older, so that the date from near the base of the core is calibrated to between 5,292 and 4,881 calendar years ago. Note how when *Sphagnum* is common, Ericaceae are rare and vice versa. A more detailed macrofossil diagram for this site—showing a wider range of taxa—is given by Davis and Wilkinson (2004).

So ecosystem function should be more 'reliable' with more species. This engineering perspective is consistent with the ideas of Lovelock and Grime described above; however, it illustrates another great problem with such theories, namely producing a sensible definition of reliability in this context. Consider a car: it essentially either takes you where you want to go or doesn't; a washing machine either washes your clothes or doesn't. So for these complex machines, defining their reliability as how long they work properly until they fail is easy. However, it is not so easy to identify the function of an ecosystem so unequivocally, so defining reliability is tricky. One approach might be to quantify how some aspect of the ecosystem (say total biomass) changes in response to imposed perturbations (like a fire, or introduction of an invasive species). It would seem humanity is perturbing a lot of ecosystems right now, sadly giving us an abundance of study material. Any younger scientist reading this book can either be dismayed at the state of the world that people their age are inheriting or be inspired that there is clearly a real importance to working in the environmental sciences over the next few decades.

4.6 Overview

Trade-offs are a fundamental aspect of biodiversity as they prevent one or two species from monopolizing the planet. As such they allow speciation, so that on any hypothetical planet with life we would expect to see a range of taxa. Well-known ecological concepts such as the niche only make sense in the context of the more fundamental idea of trade-off, as if organisms could be good at everything (supergeneralists or Darwinian demons) then the niche concept would be useless. Therefore trade-offs are a more fundamental idea than the niche or the classical mechanisms of speciation, as trade-offs are required for these other processes to operate. The resulting biodiversity will have a positive Gaian effect; that is it will tend to make an ecological community, or a planetary ecosystem, more stable than if it was composed of a smaller number of species. Biodiversity does not evolve to help stabilize the system (except in the limited but potentially important sense that taxon-poor systems may be more prone to extinction—a topic that will be returned to in chapter 12). Biodiversity is an inevitable by-product of trade-offs and other processes such as geographical isolation. One potentially important way to think about the Gaian effect of biodiversity is the idea of the 'portfolio effect' from economics, although other ideas such as Grime's 'transient species' are also important in this context. In the universe, stability is likely to be an important by-product of biodiversity which should emerge on any planet with life.

CHAPTER 5

Dispersal

5.1 Tansley's beechwoods

Woods dominated by beech trees (Figure 5.1) are a characteristic landscape feature of parts of southern England. Their colours change through the year, with in 'May the tender translucent green of the young leaves, in high summer the cool of shade beneath the heavy deep green ... [then] the glorious golden bronze of the autumn foliage' (Tansley, 1949). This description comes from a popular book on British vegetation, written in retirement by the influential plant ecologist Arthur Tansley. Beechwoods were clearly special to Tansley, and the opening of the chapter on these woods is in a far more lyrical prose style than seen elsewhere in the book, as he made the case that, 'For their loveliness alone we ought to preserve our few beechwoods with jealous care'.

Tansley was a key figure in the early history of ecology, playing an important role in developing the idea of the ecosystem in a paper published in the American journal *Ecology* in 1935. He had a strong interest in theory—wanting to understand the processes of ecology, not merely to document and describe the patterns seen in nature. I am writing this with Tansley's copy of F.O. Bower's (1930) book on *Size and Form in Plants* open in front of me. Tansley's marginalia are scattered through the book, especially 'Why?' pencilled next to several of Bower's 'actual observations of the form of plants' (Bower, 1930, p xiv). Bower had less sympathy with more theoretical approaches to science, for example describing Tansley in his correspondence as 'a rather dreamy philosopher' (Ayres, 2012). It's unlikely Frederick Bower would have had much time for my book.

English beechwoods, so beloved of Tansley, provide a nice example of why I have included 'dispersal' as one of my candidate fundamental processes (it was absent from the first edition of this book). Tansley (1949) noted that it was only recently that the fact had been fully established that the beech is native to Britain. Part of the problem appears to have been that Julius Caesar's accounts of Britain appeared to say that beech was absent from the country. However, it seems likely that Caesar had used the word '*Fagus*' to mean sweet chestnut *Castanea sativa* rather than beech (Godwin, 1975). We now know, due to pollen and other plant remains retrieved from sediments, that beech was probably the last of the main British tree species to naturally recolonize Britain after the last glaciation. It arrived in southeast England by at least 3,000 years ago after expanding across Europe from a restricted range, mainly in southern Europe, 13,000–10,000 years ago (Birks, 1989; Magri, 2008). Without the ability to disperse it would have remained trapped with a very restricted European distribution. In fact, during the last glaciation the southern English sites that now have the beechwoods Tansley so admired would have looked more like arctic tundra. Without an ability to disperse, no non-tundra plants (or other organisms) would have been able to colonize southern England as the climate warmed, and presumably many of the tundra species would have become extinct as the environment became warmer. This example suggests that without dispersal the long-term persistence of life may be impossible—this idea is developed in this chapter, first for life on an Earth-like planet and then more speculatively at the scale of solar systems and whole galaxies.

The Fundamental Processes in Ecology. Second Edition. David M. Wilkinson, Oxford University Press.
 DOI: 10.1093/oso/9780192884640.003.0005

Figure 5.1 Beech trees *Fagus sylvatica* showing 'the glorious golden bronze of autumn foliage' (Tansley, 1949, p 97), in this case in Wytham Woods, near Oxford, England. Arthur Tansley was Professor of Botany at Oxford from 1927 until his retirement in 1937 (Ayres, 2012). See also Figure 5.4.

5.2 The evolution of dispersal

Although over the long term any species that is unable to disperse may be destined to extinction there are interesting short-term evolutionary questions. Dispersal is likely to be dangerous, with many individuals not arriving safely at a suitable new site. For an organism that lives in an unstable habitat, with patches of suitable conditions appearing and disappearing all the time, such risks may make evolutionary sense. There are opportunities for any individuals that can reach fresh sites. Good examples of such an organism are the insects that spend most of their lives in fresh animal dung. But what about species that live in more stable habitats? Why take the risks of dispersal if conditions are likely to be good at your home site and unlikely to be any better at a new site? Although dispersal may be in the long-term interest of the survival of the species, if it leads to a lower breeding success for dispersing individuals it would seem unlikely that natural selection would favour it. The classic, and highly informative, theoretical discussion of this problem is provided by Hamilton and May (1977).

To understand the problem, and Hamilton and May's approach, it is useful to think about an actual example. Many plants in the daisy family (the Asteraceae) have tiny fruits which are wind dispersed with the aid of fluffy bristles referred to as a pappus (Christenhusz et al., 2018). The many species of dandelions, thistles, and similar species provide good examples (Figure 5.2) and were used

Figure 5.2 Many plants in the Asteraceae, such as dandelions *Taraxacum* spp., and other similar-looking taxa have wind-dispersed fruits. These can disperse widely, but many fail to arrive at sites suitable for germination—so such widespread dispersal seems a risky trait to have evolved.

Table 5.1 A simple illustration of why dispersal is advantageous even in a stable environment when new sites are no better than the home site and dispersal increases the chance of mortality (after Hamilton and May, 1977). Imagine a plant that grows at several different locations (in this case at seven different sites). Most plants produce seeds that don't disperse and so stay at the home site (adult plants are called 'H' and their seedlings 'h') while others produce seeds which can be widely dispersed by wind ('D' and 'd'). Each plant can produce five seeds which turn into seedlings. Imagine there is the opportunity for one of these to grow to maturity at any one site—so at sites 1, 2, 5, and 6 each 'h' seedling has a 1/5 chance of becoming an adult and reproducing. The dispersal (D) plants also produce five seedlings—one lands at its home site (4) while the others disperse. Because this is risky, half of the dispersing seeds die; however, two start to grow (at sites 3 and 7) and have a 1/6 chance of taking over their new site (note this is in addition to successfully establishing a new plant at its home site). A little thought about the implications of this table suggests that even though through dispersal the seed has a 50% mortality rate and there are no advantages to the new site over the home site, such a strategy will lead to the spread of 'D' plants compared to 'H' plants.

Sites	Adult plants	Seedlings	Seedlings	Seedlings	Seedlings	Seedlings	Seedlings
1	H	h	h	h	h	h	
2	H	h	h	h	h	h	
3	H	h	h	h	h	h	d
4	D	d					
5	H	h	h	h	h	h	
6	H	h	h	h	h	h	
7	H	h	h	h	h	h	d

by Hamilton (1996) as a real-world example of the organisms addressed by the model. The fact that the plant has reproduced successfully at the site should mean that conditions are suitable at that location; however, given the chancy nature of wind dispersal, many of its fruits will blow to unsuitable locations and not survive. The fitness of the plant can be thought of as the expected number of suitable sites to be gained by its offspring (the chance of retaining the home site plus any new sites colonized by dispersal). Hamilton and May provided mathematical models of this using a game theory approach, but started their paper with a simple conceptual model to illustrate the basic principles (Table 5.1). Both this and their more formal models showed that even in a stable habitat there are significant advantages to dispersal. An organism that can disperse still has the opportunity to maintain itself at the site where it is growing (as does a non-dispersing organism), but it also has a chance (even if it's very low) of colonizing a new site. The key point for the large-scale questions addressed in this chapter is that some level of dispersal is likely to be selected for, even in species living in very stable habitats where it might appear that not dispersing is the best strategy.

5.3 Dispersal as a key process in ecology

The example of beech and other plants dispersing in response to the large-scale climatic changes at the end of the last glaciation illustrates that dispersal is required for survival at geological timescales. This is an idea that has been understood since the realization in the nineteenth century that there had been one or more ice ages in the recent geological past which must imply widespread colonization of Britain by plants and other organisms as the climate warmed (Reid, 1899). More recently the logic of Hamilton and May's model showed that even over short timescales there will be selection pressures for some degree of dispersal for even the most sedentary of organisms. Dispersal has long been an important topic in biogeography; however, over the last few decades it has also become important in population ecology due to the rise of ideas of metapopulation ecology.

Consider an environment where suitable habitat for an organism comes in patches—these might be natural, such as a series of ponds for an aquatic species, or a result of human activity, such as a series of small woodlands forming islands of more 'natural' habitat in an agricultural landscape (Figures 5.3 and 5.4). These patches (ponds or woodlands) might

Figure 5.3 Landscape changes by humans have made many habitats patchier—in this case patches of woodland separated by agriculture near Basel in Switzerland. The idea that human modification of the landscape has often increased the patchiness of the environment, and so the importance of dispersal, has been around for decades. For example MacArthur and Wilson (1967) illustrated the reduction and fragmentation of woodland in Wisconsin, USA in the opening chapter of their important book on the theory of island biogeography. This has led to the widespread suggestion that weak dispersers are particularly at risk from human-driven landscape fragmentation. However, a review of relevant studies found that only 56% of reported tests were consistent with this prediction (Martin et al., 2023). One reason for this surprisingly low-looking figure is that classifications into 'strong' and 'weak' dispersers may be far from exact. As described in the main text, many plants, and other organisms, are described as having no obvious means of dispersal and yet they may often be dispersed by birds. So organisms assumed to be weak dispersers may be able to disperse over surprisingly large distances.

be too small to allow the long-term persistence of a particular species; however, if it can disperse between them it can (re)colonize empty patches at a rate that counteracts the effects of extinction in current patches. So any particular patch may be occupied or empty at a given point in time, but dispersal between patches means that some patches will contain a population at any point in time. For the population as a whole to become extinct the rate of patch-level extinction has to exceed the rate of recolonization. This is a metapopulation—a population of populations linked together by dispersal (Hanski, 1998, 2016; Nee, 2007). Over the last three decades or so it has become a very prominent idea in ecology; one marker of this is that the most recent edition of the leading ecology textbook by Begon and Townsend (2021) has a 37-page-long chapter on 'Movement and metapopulations'. Indeed Zhang et al. (2021) argue that in the case of spatially heterogeneous environments the important ecological idea of carrying capacity is best thought of as an emergent property that depends on the population growth and dispersal rates. Given that spatially uniform environments are likely commoner in theoretician's abstractions than in the real world, most environments are spatially heterogeneous.

The role of dispersal in metapopulations can be clearly appreciated from one of the simplest mathematical representations of the idea (Levins, 1969, although I have followed the slightly clearer nomenclature used by Begon and Townsend (2021)):

$$dp/dt = cp\,(1 - p) - ep$$

where p denotes colonized patches, c is the rate of recolonization of empty patches, and e is the rate of local extinction of colonized patches. This equation makes it easy to appreciate that persistence of the metapopulation is due to a balance between colonization of patches (cp) and extinction of occupied patches (ep). So dispersal has a central role in many aspects of modern ecological theory. Indeed Zhang et al. (2021) recently concluded that 'Dispersal, *as a fundamental ecological process*, plays an important role in shaping population dynamics, community diversity and composition, and ecosystem functioning' (my italics).

Given the importance of dispersal in ecology and also as a way of organisms responding to both past and future climate change, understanding how this dispersal happens is increasingly urgent. We are currently adding carbon to the atmosphere at unprecedented rates, with resulting climate changes that 'by the end of the century, may surpass

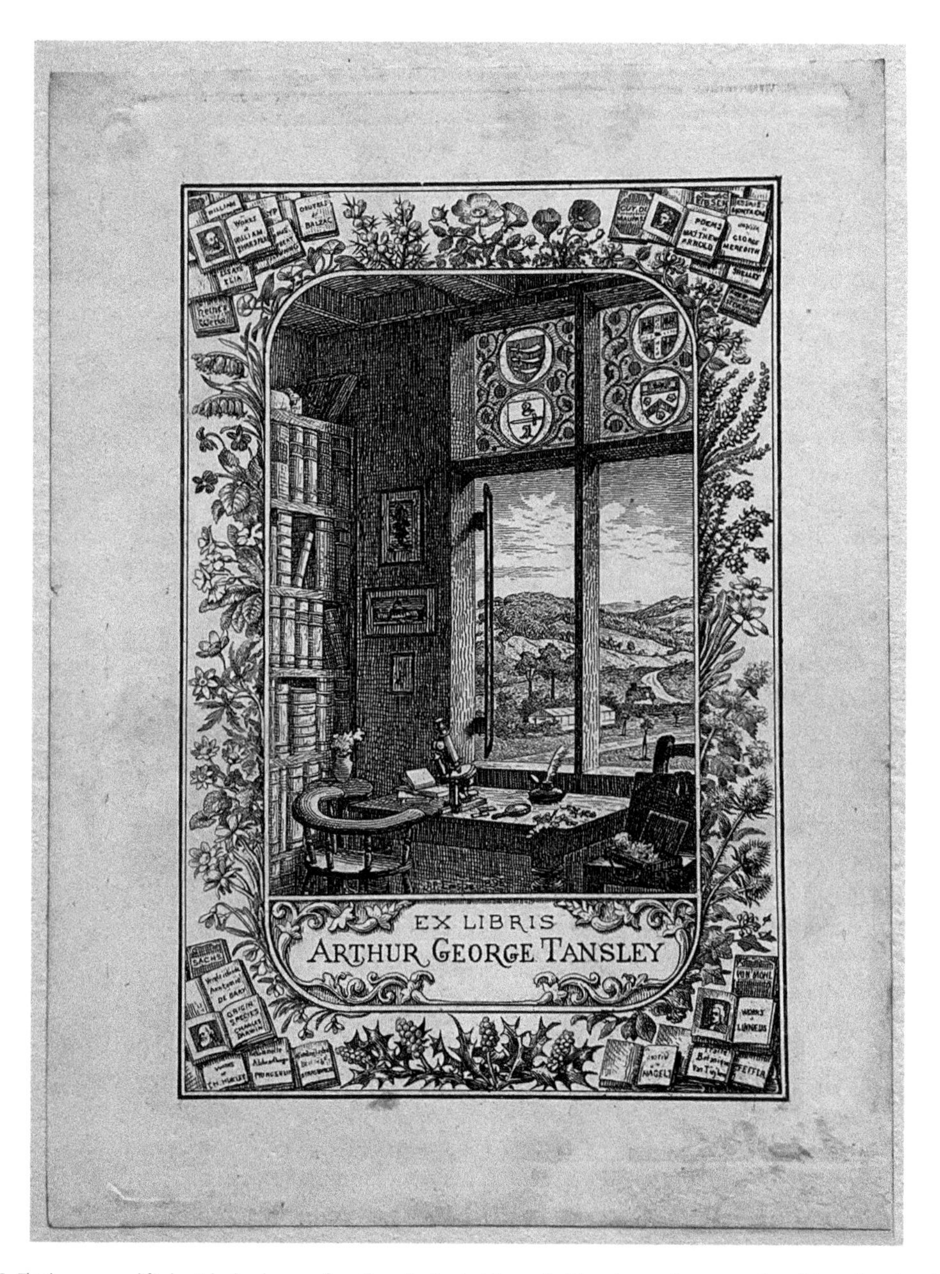

Figure 5.4 The human-modified patchy landscape of southern England at the end of the nineteenth century, viewed through an imaginary study window in Arthur Tansley's book plate—which for some years he used to indicate his ownership of a book in his library. Tansley had these book plates made when he was in his mid-twenties, around the time his interests were turning from plant anatomy (note the microscope on the table) to plant ecology—this particular example comes from his copy of Baumann (1911). The landscape is part of the South Downs, and many of the trees on the hills were beech trees (Ayres, 2012).

thresholds that triggered previous mass extinction events' (Kemp et al., 2022). Can most organisms disperse at a rate that will allow them to track these changing climates, and if so how? To try to answer such questions we need to know a lot about the natural history of dispersal.

For some types of animals the mechanism of dispersal seems obvious; for example most birds and many insects can fly. However, the majority of organisms cannot take part in such active dispersal and are instead moved by things such as wind, water, and animals—a process known as passive dispersal. Passive dispersal has been most extensively studied by botanists who have tended to try to classify plants according to dispersal classes or syndromes (e.g. Ridley, 1930; van der Pijl, 1969). The idea is that a particular syndrome, such as wind or animal dispersal, is predictable from the morphological features of a seed, spore, or similar structure. The assumption is that 'seeds' with a wing or a pappus are wind dispersed (Figure 5.2), while berries are animal dispersed. In such an approach many plants end up classified as having no obvious means of dispersal. More recently such dispersal syndromes have appeared in databases of functional traits and are, for example, used in trying to predict how plants will react to changing climate. The problem with this approach is best seen in the context of animal dispersal, where many 'seeds' with adaptations to wind dispersal, or no obvious dispersal adaptations at all, may often be dispersed by animals in addition to other mechanisms (Wilkinson, 1997; Green et al., 2022). This matters because birds may disperse them over much longer distances than seem likely for a 'wind-dispersed' seed.

So even such long-established research areas as plant dispersal are clearly in need of more work, which is particularly urgent in the context of changing climates (Figure 5.3). We know much less about the dispersal of microorganisms; however, their small size and large numbers have led some workers to suggest that they can disperse very widely indeed (e.g. Finlay, 2002). Size is likely important, with both empirical analysis (Yang et al., 2010; Lara et al., 2011) and modelling work (Wilkinson et al., 2012) suggesting that larger microbes (20 μm upward) become increasingly less likely to be globally dispersed. Because of the patterns of global air circulation it is particularly difficult for microbes to be wind dispersed between hemispheres, so possibly dispersal by migrating birds may be important in this case as many major migration routes cross the equator (Decloitre, 1953; Wilkinson et al., 2012). Theoretical studies of the extent to which organisms may be able to respond to changing climates by dispersal require good data, which can only come from empirical studies. We still have much to learn, but with pressing global changes we urgently need good answers.

Part of what makes dispersal challenging to study is that freak events may be very important. Consider the dandelion fruits used as an example above. They look highly selected for travel on the wind; however, in dense vegetation likely almost all of them will travel only a short distance. Maybe just one might be swept up on an updraft and carried several kilometres before it falls back down to earth, and perhaps another one might fall in some mud and then be splashed onto the fur of a wandering mammal, being carried tens of kilometres before the fur dries out and the fruit falls to earth. So we can be misled on plant dispersal if we only study what happens to seeds on average; we really have to follow all of them.

5.4 Dispersal at an astronomical scale?

As this book uses an astrobiological thought experiment as a way of thinking about terrestrial ecology, it is of interest to briefly consider ideas on dispersal at an astronomical scale. The idea that life may not be restricted to a single planet like Earth but may be able to move between planets—or even between solar systems—has a long history. This idea, often called panspermia, started to attract more scientific interest during the 1830s as it was realized that some meteorites contained organic (i.e. carbon-containing, not necessarily biological) compounds (Nicholson, 2009; Mitton, 2022). The idea seems more likely within a solar system than at a galactic scale, so I will mainly consider transfer between planets within a solar system, and only briefly comment on the more speculative case of potential dispersal between solar systems or even between galaxies.

We know that rocks can move between planets within a solar system, as rocks from Mars can be found on Earth (most of the discussion in this paragraph is based on Nicholson, 2009). What is the probability of a microorganism successfully dispersing from one planet to another (P_{DIS})? By breaking down this transfer into a number of independent factors we can write an equation that at least makes explicit the things we would need to know to calculate such a probability (remember the multiplication of probability rule; assuming all the component probabilities listed below are independent, they should be multiplied together to give P_{DIS}, with these multiplications denoted by '*' to prevent confusion between the letter 'x' and the common multiplication symbol). The following equation describes the situation where an object from space impacts a planet with life and ejects fragments of the planetary surface into space:

$$P_{DIS} = P_{BIZ}*P_{EE}*P_{SL}*P_{SS}*P_{SE}*P_{SI}*P_{REL}*P_{SP}$$

where

P_{BIZ} = probability that an object impacting a planet strikes a biologically inhabited area.

P_{EE} = probability that rocks ejected into space contain endolithic microbes (see below and Figure 5.5).

P_{SL} = probability that these microbes survive being launched into space.

P_{SS} = probability of surviving space transit (briefly discussed below).

P_{SE} = probability of surviving entry through the recipient planet's atmosphere.

P_{SI} = probability of surviving impact on the new planet's surface.

P_{REL} = probability that the microbes can escape from the interior of their rock.

P_{SP} = probability of survival and population growth on the new planet.

While this equation is useful when it comes to being explicit about the various factors we need to understand to say something about the likelihood of interplanetary panspermia, it's not necessarily easy to use it to calculate a realistic value for P_{DIS}. It doesn't need much thought to realize that some of these probabilities will be very hard to assign actual values. Indeed Nicholson (2009) pointed out that depending on the best guesses and approximations used, different research groups have used this type of approach to conclude that successful microbial dispersal between planets was anything from 'highly probable' to 'highly improbable'. However,

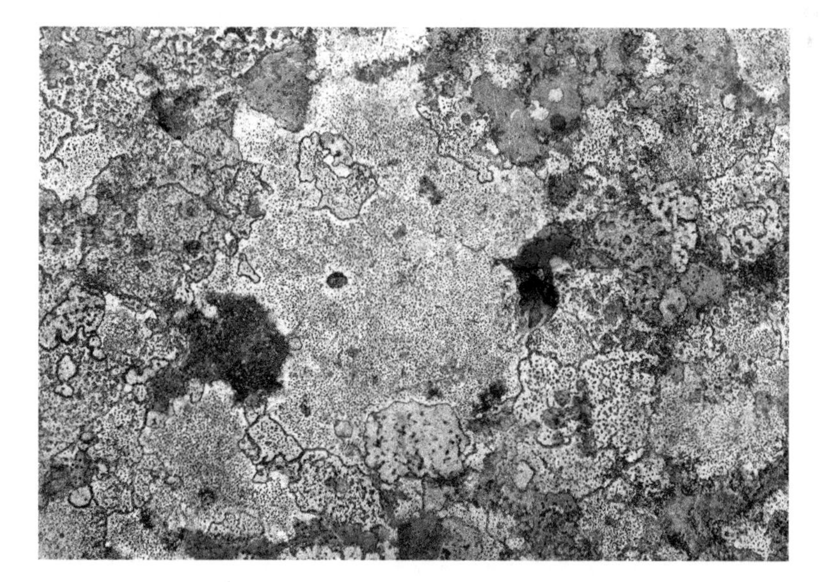

Figure 5.5 Close-up of endolithic lichens at Malham Cove, Yorkshire, in northern England. In the case of lichens, because of their requirement for light they have to live in the surface of rocks. Other non-photosynthetic endolithic microbes can live deep within rocks. Over 100 species of lichen have been recorded growing on or in the limestone within a few kilometres of this site—many of which are largely endolithic (Seaward and Pentecost, 2001). The photograph shows approximately 15 × 10 cm of rock surface. See also the discussion on long-term survival of microbes in rocks in chapter 6 (Box 6.1).

some of these component probabilities are more amenable to being quantified than others. Considering the more biological component P_{SS}, experiments have shown that various bacteria and even lichens (e.g. Backhaus et al., 2019) can survive in a dormant state in experimental simulations of Martian conditions, or even actual exposure to space. Many microbes, and even some lichens (Figure 5.5), can live within rocks—a lifestyle referred to as endolithic. This will give them some protection and they are usually assumed to be the most likely types of organisms to be transferred between planets.

Milton de Souza Mendonça Jr. (2014) has taken these ideas further, applying ideas of metapopulations at the scale of a whole solar system. Potentially life can survive longer as a metapopulation spread across multiple planets than it could if restricted to just a single planet—for example moving between planets as conditions change over time causing the extinction of some patches (planets). There is also the possibility of applying similar ideas at a galactic scale—a typical galaxy such as our own contains around 10^{11} stars, and roughly a trillion galaxies can be seen over the expanse of the night sky (Thorne and Blandford, 2017). This is a huge scale to apply ideas of spatial ecology, and any such applications must be extraordinarily speculative. Obviously compared with dispersal within a solar system there is much more uncertainty about microbes surviving dispersal between solar systems, and even more for the vast distances between galaxies (Adams and Napier, 2022)! If panspermia within a galaxy is possible then it appears much more likely in a galaxy's central region where the density of stars is much greater (Gobat et al., 2021). For panspermia across large interstellar distances there is the intriguing, if even more speculative, idea of directed panspermia—where an intelligent life form deliberately uses space craft to disperse microorganisms to other planets. The logic is it would be technically much easier to send microbes over very long distances than the intelligent life form itself (Crick, 1982). This idea was developed by Francis Crick in collaboration with Leslie Orgel, although Crick's most important biographer suggests that Crick didn't really believe the idea but 'wanted to stimulate discussion' (Olby, 2009). Indeed all such discussions about dispersal at scales larger than solar systems are currently very speculative, existing in a fascinating borderland between science and science fiction.

5.5 Overview

Unless an organism's ancestor evolved at precisely the location you find it, then it, or its ancestors, must have dispersed from elsewhere. On longer timescales the environment at a particular location will have been unsuitable at some point in the past (e.g. for a terrestrial species it may have been under either liquid or solid water) so dispersal should be fundamental to the ecology of any planet. The ideas developed by Hamilton and May also suggest that dispersal will be key to the success of any organism—however sedentary it appears or however stable its environment. Dispersal also became central to ideas of metapopulations in ecology during the latter part of the twentieth century. Such metapopulation approaches can be applied at astronomical scales, albeit speculatively, in considering the likelihood and potential importance of panspermia. The arguments developed in this chapter suggest that at a planetary level dispersal would have a positive Gaian effect, and that this may also be the case at the scale of solar systems or possibly even galaxies.

CHAPTER 6

Ecological hypercycles: covering a planet with life

6.1 Darwin's earth worms

Charles Darwin's final book addressed *The Formation of Vegetable Mould Through the Action of Worms, with Observations on Their Habits* (Darwin, 1881). In it he describes the feeding habits of earth worms, writing: 'They swallow an enormous quantity of earth, out of which they extract any digestible matter ... They also consume a large number of half decayed leaves of all kinds' (Darwin, 1881, p 35). He also describes the effects of earth worms on the soil; for example he discusses the amount of soil overlying several British archaeological sites which he largely attributes to the action of worms over the generations (Figure 6.1). Importantly he describes how this soil formation by worms improves growing conditions for plants, writing: 'Worms prepare the ground in an excellent manner for the growth of fibrous-rooted plants and seedlings of all kinds' (Darwin 1881, p 309).

The science historian Eileen Crist has argued that Darwin's worm book, although not an anticipation of Gaia, can be 'read, without strain, as a geophysiological thesis' (Crist, 2004, p 165). She points to the emphasis on how worms modify their environment in a way that is beneficial to future generations of worms and an implicit description of the positive and negative feedback loops between the worms and their biotic and abiotic environments. For these same reasons Darwin's worm book also came close to describing the idea of an ecological hypercycle.

6.2 Hypercycles in ecology

Darwin's description of earth worm ecology could be summarized by the following relationships:

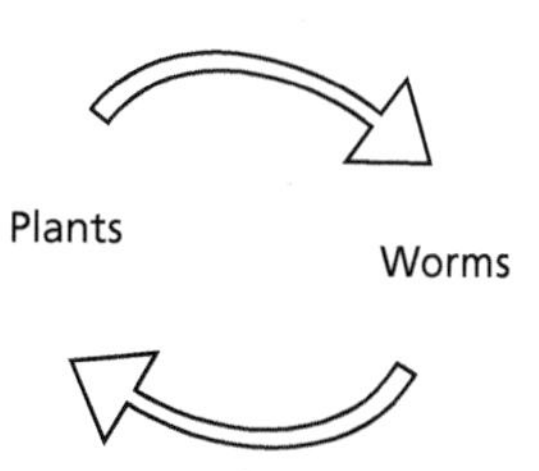

where the arrows show a positive effect of one group on the other. Although the following summary would appear obvious to a twenty-first century reader it is never explicitly described by Darwin.

This diagram shows a positive feedback loop and is an example of a simple hypercycle (Eigen and Schuster, 1977; Maynard Smith and Szathmáry, 1995), where the rate of replication of each group (plants and worms) is an increasing function of the size of the preceding group. So more plants means more worms (more food), while more worms means more plants ('better' soil for growth). One of the main applications of hypercycles in biology has been in considering the autocatalytic effects in cycles of self-replicating molecules during the origin of life; however, they are clearly also relevant to ecosystems (Maynard Smith and Szathmary,

The Fundamental Processes in Ecology. Second Edition. David M. Wilkinson, Oxford University Press.
 DOI: 10.1093/oso/9780192884640.003.0006

Grass: Nineteenth-century ground level.

'Mould, 20 inches thick' (50 cm).
—this was described as an 'unusual thickness' of mould for this site.

'Rubble with broken tiles, 4 inches thick' (10 cm).

'Black decayed wood, in thickest part 6 inches thick' (15 cm).

'Gravel'— Darwin was unsure if this was natural or archaeological in origin.

Figure 6.1 Diagrammatic representation of a soil profile from nineteenth-century excavations at the Roman site of Silchester in southern England. This was a large Roman town which was extensively excavated during the 1800s, partly because of the accessibility of the remains, as there had been relatively little post-Roman building on the site (de la Bédoyére, 2002). These data were supplied to Charles Darwin by the Rev. J.G. Joyce, one of the excavators. By 'Mould' Darwin meant good-quality soil which he assumed had been processed by passing through the guts of worms. This diagram is based on Figure 11 in Darwin (1881), with measurements in centimetres and other annotations added. Text in quotation marks is taken directly from Darwin's diagram or his accompanying text. Darwin used this and several similar archaeological studies to illustrate the rate of soil formation, over a timescale of less than 2,000 years, which he attributed to earth worm activity. These studies showed biology having a significant effect on its physical environment. In December 1957 the archaeological journal *Antiquity* had a special issue to mark 100 years since the publication of *On the Origin of Species* and the influence of evolutionary ideas on archaeology. This also included a paper on 'worms and weathering', drawing archaeologists' attention to Darwin's worm research—'fascinating work still largely unknown to archaeologists'—and to subsequent work on the effect of earth worms on archaeological sites (Atkinson, 1957).

1995: Cazzolla Gatti et al., 2018). For example, the dung-based system described in section 3.2 can be represented as such a cycle.

Feedbacks of this type are crucial to an Earth systems view of ecology, and yet they are still often overlooked by ecologists working at smaller scales (Pausas and Bond, 2022). Once again it seems Darwin was very much ahead of the game in focusing on such feedbacks in his last book. If the way the subject is presented in the textbooks is any guide, many ecologists have still not caught up. Such ecological hypercycles don't just apply to worms and plants; another simple ecological hypercycle looks like this:

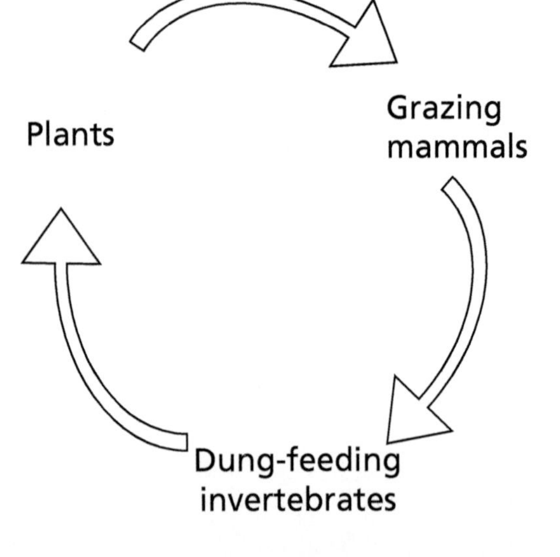

In the example above, if plants increase then there is more food for the herbivorous mammals (be they deer, antelopes, or elephants) that feed on them. If the mammals increase in numbers, then there is more food for the guild of dung-feeding invertebrates; and an increase in the numbers of these invertebrates speeds up the decomposition of the dung, so making more nutrients available for plants. Grazing lawns in African savannah systems are instructive examples (Figure 6.2). These are areas of short grass maintained by high grazing pressure, and are considered to be created by positive feedbacks between grazing animals and plants. The grazers maintain short nutritious grass through the effects of their grazing suppressing other plant species, along with positive effects from nutrient input from their dung (Cromsigt et al., 2008; Shorrocks and Bates, 2015). Key species involved in creating these lawns include white rhinoceros (Figure 6.3), hippopotamus *Hippopotamus amphibius*, blue wildebeest *Connochaetes taurinus*, impala *Aepyceros melampus*, and warthog *Phacochoerus africanus* (Bond, 2019).

One of the difficulties with such ecological hypercycles is that they are usually a great simplification of the real situation, as in reality additional arrows are often needed. Consider African grazing lawns: the positive feedbacks between grazers and nutrient-rich short grasses make a neat simple story—but fire is also involved (Midgely et al., 2010; Donaldson et al., 2017). Grazers and fires are not just two separate processes affecting these African

Figure 6.2 A grazing lawn in Kruger National Park, South Africa, photographed in July 2014 during the dry season. As Bond (2019) notes, 'though highly productive in the wet season, grazing lawns epitomize "brown world" in the dry season when the grasses are dried up'. However, although there looks to be a lot of bare earth in the photograph, our measurements showed only 64% was completely bare soil. Also note the dung of at least two species of herbivorous mammal in the foreground and the gun leaning against a shrub—this latter point illustrates another aspect of grazing lawns, as the short vegetation makes it hard for any predators to sneak up on the herbivores (hence our game guard felt safe putting his gun down).

Figure 6.3 White rhinoceros *Ceratotherium simum*, a key species in the formation of African grazing lawns, especially in the more productive savannah systems (Bond, 2019). This individual was photographed in Kruger National Park and is part of the reintroduced population studied by Cromsigt and te Beest (2014), who found that the rhinos had had a significant effect on the extent of grazing lawns in the park.

savannahs but interact synergistically too. Burnt areas can attract grazing animals and help create grazing lawns. For example, in the experiments of Donaldson et al. (2017) in Kruger National Park, small repeated burns were effective in creating new grazing lawns. However, once the grazing lawns exist, their short vegetation reduces the incidence of fire. Pausas and Bond (2018) have generalized from examples such as these to suggest that the long-standing idea that climate and soils are the only major factors controlling large-scale patterns in vegetation is often inadequate, and we need to embrace 'fire and large mammal herbivory as additional key factors in explaining the ecology and evolution of world vegetation'.

Although a simplification, the idea that life can have an autocatalytic quality is of great importance in understanding the ecological history of a hypothetical planet (or real planet, such as Earth). Once autotrophs and decomposers have evolved, as discussed in chapter 3, then the simple hypercycle **autotroph ↔ decomposers** will exist. Note that this is really just a generalized version of the **plant ↔ earth worm** hypercycle described above. This has implications for the spread of life on any planet on which it evolves—or on which it arrives if you favour panspermia rather than an *in situ* origin of life. In addition Cazzolla Gatti et al. (2018) have argued that such autocatalytic processes may have a role in creating biodiversity through a 'niche emergence' process. For example, once you have trees then that creates new niches for tree-nesting animals or epiphytic plants.

6.3 Covering a planet with life

These autocatalytic hypercycles, such as the **autotroph ↔ decomposer** example described above, could quickly lead to life covering much of a planet, as the presence of life catalyses the production of more life. Indeed, Nisbet and Fowler (2003, p 22) suggest that 'on a geological timescale, once the first cell had replicated, all habitats on the planet would be immediately filled', although on Earth there was probably a considerable lag before life started to appear in terrestrial habitats. Why should this be the case? As well as hypercycles, another mechanism is the well-known potential for geometric growth in unconstrained populations (when the net reproductive rate $[R_0] > 1$) which also leads to the same conclusion: namely life quickly covering the planet. Equations of the form:

$$N_{t+1} = R_0 N_t$$

are common in ecology textbooks (where N_t is population at time t and N_{t+1} is population at the later time of t+1), so once life starts reproducing there is the potential for it to quickly reach huge population sizes. This large amount of life can then be subdivided into different species by the processes described in previous chapters. Similar arguments (along with those discussed in chapter 7) are also the reason why Lovelock (2000a) argued that planets will tend to have either lots of life or no life at all.

This conclusion, of life covering at least the aquatic part of a planet in a geological instant, clearly requires the habitat for life to be widespread. On Earth it is usually assumed that early life could survive in aquatic habitats and since the planet had widespread oceans (although some of these may have been frozen) it could quickly spread (Nisbet and Sleep, 2001; Nisbet and Fowler, 2003). This is important in the context of the idea of Gaian bottlenecks—the idea that once life has evolved on a planet, if it is to survive it needs to spread to the point where it can start to play a role in the planetary feedbacks that help maintain habitability (Chopra and Lineweaver, 2016). This is an idea I will return to in more detail in chapter 7.

6.4 Why would persistent restricted ecologies be unlikely?

An alternative possibility is that life may survive on a planet in only a few locations. There are at least two main reasons why this is unlikely: one is because of environmental stochasticity tending to cause extinction (described below), while the other is that it would be unable to play an important role in the 'physiology' of its planet (the subject of chapter 7) and as such be unable to contribute to planetary regulation. The first of these reasons will be familiar to most ecologists as it is effectively part of the 'small population paradigm' of conservation biology applied at a planetary scale. One of the reasons why conservationists worry about small populations is they are prone to extinction brought about by demographic and environmental stochasticity (Lande, 1993; Caughley, 1994; Simberloff, 1998a—see Hambler and Canney, 2013 for a textbook account).

Demographic stochasticity is when chance fluctuations in demographic parameters (e.g. birth and death rates, sex ratio, population growth rate) in a small population can lead to extinction. Consider sex ratio in the simple situation where each generation of an organism does not overlap—for example the adults produce eggs or seeds and then die off before the next generation is ready to breed (many insects and annual plants show this behaviour). It is intuitively obvious that if there was only one member of the next generation then it must either be male or female, giving the trivial result that the probability of all members of that generation being the same sex would be 1 (i.e. 100%) if the generation size was 1. If this species has an equal probability of producing offspring of either sex and there were two individuals in the next generation then the probability of them both being the same sex would be 0.5 (logically the probability of the second individual being the same sex as the first must be 0.5). It is possible to calculate probabilities for more complex cases using the following relationship (Simberloff, 1998a):

$$\text{Probability of all individuals being the same sex} = 2^{1-N}$$

where N is the number of individuals in a given generation. The results of applying this relationship to several small population sizes are shown in Table 6.1. As can be seen, at very small population sizes (e.g. 5) there is a real possibility of all individuals being of the same sex by chance. By the time the population size is 10 individuals this is quite unlikely (but would still be expected to occur once or twice in every 1,000 cases), while by the time the population consists of 100 individuals it is exceedingly unlikely. While this particular example only applies to sexually reproducing organisms, these results are typical for demographic stochasticity, in that these processes apply mainly to very rare organisms. As such these ideas are much less relevant to microbial ecology than they are to large multicellular organisms such as the Californian condor *Gymnogyps californianus*. For example, while there were only around 20 California condors in the wild in the early 1980s before they were taken into captivity (del Hoya et al., 1994), there can be around 4×10^4 cells/ml of the marine prokaryote *Prochlorococcus* spp. (Whitman et al., 1998) and even the much larger testate amoebae species can reach densities

Table 6.1 Probability of all individuals in a generation being of the same sex for a species that produces males and females with equal probability and has non-overlapping generations.

Number of individuals	Probability all of one sex
2	0.5
5	0.062
10	1.9×10^{-3}
100	1.6×10^{-30}

of 2,300 individuals/cm^3 in peatland soils (Tolonen et al., 1992). So the very low population sizes of relevance to demographic stochasticity are not usually relevant to microbes, and as such not relevant to the early history of life on a planet, apart from a possibly exceedingly brief period at the very start of life when for a geological instant, population sizes may have been almost non-existent.

Ideas of environmental stochasticity and catastrophes are very different, being potentially much more relevant to arguments that persistent restricted ecologies don't survive for geological periods of time. Good illustrations of the importance of these processes in conventional conservation biology are given by several bird species. An example are the frigatebirds (Fregatidae); there are currently five species, three of which have quite widespread distributions in the tropics, while two are restricted to breeding on a single island: the Ascension frigatebird *Fregata aquila* and the Christmas frigatebird *F. andrewsi* (del Hoyo et al., 1992). The population of mature female Ascension frigatebirds (Figure 6.4) is estimated to be around 9,350 birds, and this number has likely not changed significantly since the late 1950s (Ratcliffe et al., 2008). While this is quite a small population it is still much too large to be easily put at risk by demographic stochasticity. However, it only breeds on Ascension Island and in recent decades its breeding has largely been restricted to the small Boatswain Bird Islet, off the east of the main island, because of feral cat predation (these cats have now been exterminated and the first frigate birds bred on the main island again in 2012). As such the whole population of birds could be wiped out by a single catastrophe, such as unusual weather events or outbreaks of disease.

Figure 6.4 Ascension frigatebirds *Fregata aquila* only breed on Ascension Island in the tropical South Atlantic. Such relatively rare species can be prone to extinction through environmental stochasticity.

Another well-known example of an island bird that did suffer extinction after humans arrived on its island home is the dodo *Ralpus cucullatus*. This was only found on Mauritius, so once it was extinct on this one island it was globally extinct; therefore its very limited range was a key factor in its extinction. A particularly good example of environmental stochasticity is the demise of another flightless bird, the great auk *Alca impennis*. This penguin-like seabird used to be relatively common in the North Atlantic but became easy prey to human hunters as their remote breeding colonies became more accessible. One of its last main breeding colonies was the island of Geirfuglasker off the coast of Iceland and the catastrophic destruction of this island by volcanic activity in 1830 hastened the demise of what was by then a rare bird (Fuller, 2000); earlier in its history when it was more widely distributed this disaster would have been much less of a threat to the survival of the whole species.

As these avian examples demonstrate, organisms that have very restricted distribution are vulnerable to catastrophes—both physical such as volcanic eruptions and biological such as disease outbreaks (although disease may seldom be enough to cause extinction by itself: see section 3.4). This is an important reason why life is only likely to show long-term survival if it covers much of a planet, although it should be pointed out that in Lande's (1993) models, populations with high growth rates were surprisingly resistant to environmental stochasticity, so even geographically restricted microbial populations may survive all but the most dramatic planetary catastrophes.

6.5 The end of life on a planet

If the idea that spatially restricted ecologies are unlikely to survive is true, then it raises interesting questions about the end of life on a planet. A good way to think about this problem is to use Mars as an example and make the big assumption that this planet had widespread microbial life in the past. Today Mars gives a lifeless impression; its atmosphere—very different from that of Earth—shows that biology must at least be exceedingly rare. Before spacecraft had ever landed on the planet, James Lovelock and Lynn Margulis (Lovelock and Margulis, 1974; Margulis and Lovelock, 1974) had used arguments about the nature of the Martian atmosphere to predict the lack of life. However, if life had been widespread at some point in the past then there would have been large numbers of microbial cells—for example there are currently an estimated 10^{30} prokaryote cells on Earth (Whitman et al., 1998; Bar-On et al., 2018). It may be very difficult to bring about the extinction of such a colossal number of microorganisms by processes other than the catastrophic destruction of the whole planet. Indeed, analogies from Earth-based microbiology suggest that it will probably take a long time for all microbial cells to die. On Earth large numbers of microbes live in the surface of rocks in seemingly inhospitable deserts; this includes both hot dry sites such as in the Californian desert (Kuhlman et al., 2005) and cold dry sites such as in the Antarctic dry valleys (Wynn-Williams, 1996). Both of these habitats have attracted the interest of researchers curious about the possibility of life on Mars. Earth-based microbiology also suggests that microorganisms (especially prokaryotes) are able to survive in a dormant state for a very long time; some papers have suggested survival for up to 600 million years, although these studies are very controversial and rather lower figures are probably more realistic (Box 6.1).

In theory microbial communities should survive better in a dormant state on a dying planet, as they will be less likely to be attacked by other organisms than on an ecologically highly active planet (such as Earth). An ability to remain dormant, potentially for hundreds of millions of years, along with an ability for small numbers of cells to survive generally worsening conditions in protective microhabitats such as within rocks would lead to life having a long 'tail' on a dying planet, petering out over hundreds of millions of years. However, there is a problem with

Box 6.1 Survival of dormant microorganisms

How long can bacteria and other microbes survive in a dormant state? There is a long history of studies that claim to have isolated viable microbes—usually bacteria—from geological sediments. All such studies have to confront two very difficult problems: first, avoiding contamination with modern microbes; and second, determining that the microbes were incorporated into these sediments at the time of formation, rather than at a later date. In addition, the accurate dating of the sediments can prove difficult. This is not an easy area of research, although the results are potentially fascinating!

Kennedy et al. (1994) tabulated the published evidence for long-lived microorganisms. When data for claimed survival of 1 million years or more is plotted against publication date, an interesting pattern emerges (Box Figure 6.1). With the exception of the pioneering work of Lipman (1928), little was published until the middle of the twentieth century, when a series of claims for microbial survival of hundreds of millions of years were published, several suggesting survival from the late Proterozoic. Such extreme claims vanished from the literature in the later twentieth century; more recently papers typically argued for survival of between 1 and 200 million years. There are two main explanations for this lack of older ages since the 1960s. One is that improved experimental methods may have greatly reduced the chance of contamination; the other is that the more robust refereeing of papers has made it far harder to publish such claims (even if they are correct!).

Box 6.1 *Continued*

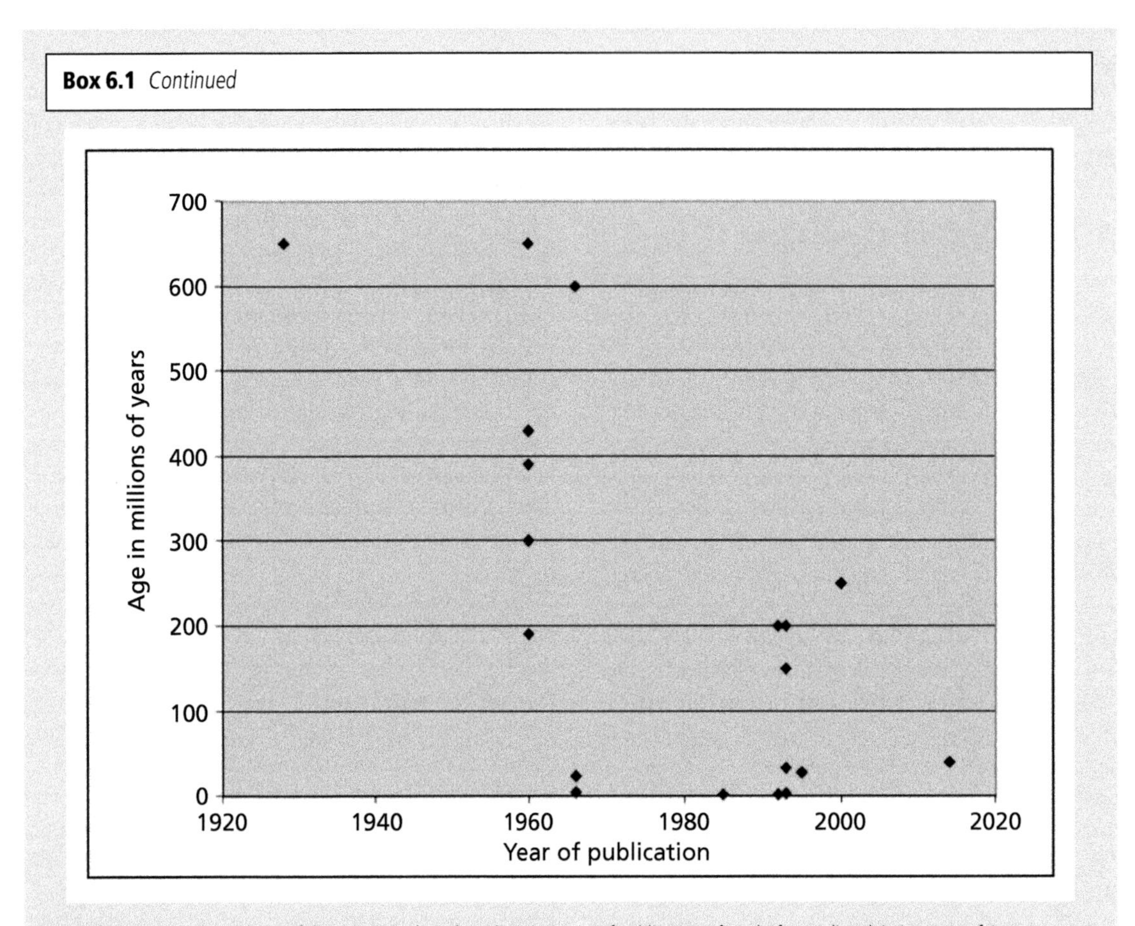

Box Figure 6.1 Claimed age of dormant microbes plotted against year of publication of study, for studies claiming ages of 1 million years or more published during the twentieth century. Data taken from Table 1 of Kennedy et al. (1994), with addition of studies of Cano and Borucki (1995), Vreeland et al. (2000), and Jaakkola et al. (2014).

For example, Vreeland et al. (2000) published evidence for 250-million-year-old bacteria from an inclusion in a salt crystal; however, even this detailed study was challenged, with doubts over the age of the salt inclusion (Hazen and Roedder, 2001). While these controversial claims of survival for hundreds of millions of years are the most relevant to arguments for the long tail of life on a dying planet, records of survival for hundreds or thousands of years are more common and usually less controversial (Postgate, 1990; Kenney et al., 1994; Miskin et al., 1998); however, these are still subject to the same problems of contamination and dating.

An alternative approach is to try to directly calculate the survival half-life for a population of dormant bacteria. For example, some bacteria can survive radiation doses of up to 2,000–3,000 Gy, with some endospores surviving much higher levels (Madigan et al., 2012); however, under most conditions on Earth this could be achieved from background radiation in a few million years. Similarly, a dormant microbe would be subject to a slow death from the simple diffusion of its cell contents over time. Considerations such as these make lifespans of hundreds of millions of years *in a dormant state* appear unlikely. However, these are probabilistic processes (very like the idea of radioactive half-life); since microbial populations can be unimaginably large, a lucky few cells could survive for much longer than the vast bulk of the population. In this case Mars becomes interesting as water seems to be present, although ideas about the amount of water in the planet's past have a long and controversial history (Morton, 2003). If extensive aquatic habitats were available in the past, could there have been life, and if so could a small living biomass still survive?

such arguments; if the cells are restricted to surviving in a dormant condition, they will be unable to repair the unavoidable damage they will accumulate over time, and so eventually die out, although it is very difficult to attach numbers to such a process (Box 6.1). Certainly, the potential 'long tail' to a distribution of microbial lifespans would make it impossible to identify the moment of death of all life on a planet. As Oliver Morton (2003, p 297) wrote of Mars, 'if you search for individual living organisms, it is almost impossible to answer in the negative; to say categorically that there is no living thing on Mars would require knowing the planet with an intimacy that is hard to imagine'. However, long before the last cell died a planet could be declared ecologically dead, its biogeochemical cycles having become effectively just geochemical cycles, of no professional interest to the ecologist. The exceptions to a long time span for the extinction of life on a planet would be catastrophic events, such as the massive asteroid or cometry impacts suffered by Earth in its early history or a potentially lethal flux of cosmic rays from a nearby supernova—here nearby may mean around 30 light years or less (Narlikar, 1999). Even in these cases it may be very difficult to kill every last microbe living deep within a planet's rocks.

6.6 Overview

Ecological hypercycles are autocatalytic processes by which different organisms (or guilds) improve each other's environments, a good example being autotrophs producing material of use to decomposers and the decomposers releasing nutrients which are reused by the autotrophs. Such an autocatalytic process clearly has a positive Gaian effect and is one of several reasons for expecting life to quickly cover most of a planet's surface once it has evolved. Environmental stochasticity makes it unlikely that ecologies restricted to a small area of a planet will survive for a geological period of time. However, once widespread, life may have a long drawn out end on a dying planet, with some cells surviving for at least millions of years after life last played any role in the planet's physiology (that is after biogeochemical cycles returned to being mere geochemical cycles).

CHAPTER 7

Merging of organismal and ecological physiology

7.1 From beavers to planetary ecology

The beaver is a classic example of an organism that modifies its environment for its own benefit. There are only two extant species of beaver in the genus *Castor*: the North American beaver *C. canadensis* and the Eurasian beaver *C. fiber*. Both species modify their environments through the building of dams and other engineering works, with similar behaviours and densities of dams recorded from both North America and Western Europe, although there is some suggestion that the European species may be less prone to large-scale engineering works (Rosell et al., 2005; Busher, 2016). The activities of beavers affect streams in at least six distinct ways (Coles, 2001): 1) altering channel dimensions; 2) favouring the development of multiple channels; 3) changing water velocities; 4) increasing the volume of water in stream channels; 5) changing the pattern of sedimentation; and 6) increasing organic input to streams (c.f. Hodkinson, 1975). Clearly the activity of beavers doesn't just affect the plants they eat but also has important effects on their physical and biological environment (Figure 7.1). Because of these effects there is now a big interest in reintroducing beavers to places, such as Britain, where they became extinct, to contribute to both hydrology (such as flood control) and biodiversity, through creating a more varied range of habitats (Wilson et al., 2020).

The extent to which beavers construct these engineering works depends upon the local hydrology and other factors. They usually build dams where the water is shallow, and by raising the water level they produce ponds that allow them both to swim to their food sources (parts of trees) and store food underwater. In addition, they can create canals that allow them to float branches and so more easily move them around (Rosell et al., 2005; Busher, 2016). Therefore by modifying their environment—'ecological engineering' in the terminology of Jones et al. (1994, 1997)—they provide themselves with a better environment. However, their engineering also modifies the environment for other species. In general, species that like wetter conditions can do well as a by-product of beaver activity, especially if they prefer still rather than flowing water (for example I have watched large numbers of dragonflies breeding in beaver ponds). Those that like dryer conditions or flowing water will often suffer as a result of the beavers' ecological engineering (e.g. you often see dead trees standing in beaver ponds that have been killed by the raised water levels). One example of the potential effect of beavers on other species is the suggestion that beavers have a positive effect on the environment for otters *Luttra* spp. in both Europe and North America. Beaver ponds provide a good hunting environment for otters and in winter otters also benefit from the beavers' habit of maintaining ice-free areas as breathing holes and for access to the water. This had led to the suggestion that late-twentieth-century increases in otter numbers in part of North America may be due to increased populations of beavers (Rosell et al., 2005).

Chapter 6 argued that life will very quickly colonize a whole planet, and an inevitable consequence of this is a large biomass of life on any inhabited world; as such, life not only will influence its local environment but also has the potential to

The Fundamental Processes in Ecology. Second Edition. David M. Wilkinson, Oxford University Press.
 DOI: 10.1093/oso/9780192884640.003.0007

Figure 7.1 Beavers modify the vegetation both by creating flooding and importantly by felling trees for food, as seen here. Although renowned for damming watercourses this is not always the case, and if a natural watercourse is deep enough they may not need to construct dams (Wilson et al., 2020). In the case shown in the photo, a small river running into Lac Léman in Switzerland, dams were not required but their feeding activities still modified the riverside woodland habitat.

cause important effects at the planetary scale. These influences could be through direct adaptations to modify the environment for its own benefit (e.g. a beaver dam) or accidentally through by-products (e.g. carbon dioxide from respiration or oxygen from photosynthesis). As the intention of this book is to focus primarily on concepts, I will first discuss a simple but informative model illustrating the potential for life to influence planetary ecologies, before discussing some of the mechanisms by which this can happen. I will then argue that such an approach forces us to give biomass—rather than species richness—much more attention than has traditionally been the case in ecology, and further argue that the mass ratio theory should be at the centre of planetary, as well as ecosystem, ecology.

7.2 Daisyworld

Open a textbook on mathematical ecology (e.g. Case, 2000) and along with the logistic growth curve, you will find that the Lotka–Volterra models of interacting populations have formed a starting point for much of twentieth-century theoretical ecology. These models are typical of much of mathematical ecology in modelling the interaction between organisms without considering feedback between the organism and its abiotic environment. Although the majority of modern ecologists probably only know of Alfred Lotka through his eponymous equation, it was not typical of his approach to ecological theory. Lotka—working in the 1920s—regarded the organic and the abiotic aspects of the world as a single system in which it was impossible to understand the working of any part of the system without considering the whole (Kingsland, 1995). This led him to suggest that modelling the whole system (biotic and abiotic) would prove simpler than attempting to understand an unrealistically isolated fragment—an approach sympathetic to that taken in this book. However, mainstream theoretical ecology has usually proceeded by considering organisms as isolated systems. While this may be a reasonable simplification for modelling *some* ecological systems of limited spatial and temporal scale, it is certainly not appropriate for planetary-scale systems.

Since the early 1980s an alternative approach to ecological modelling has developed that is much closer to Lotka's philosophy and illustrates the potential global importance of the presence of life on a planet. The origin of Daisyworld models lies in an attempt by James Lovelock to counter claims of teleology levelled against his Gaia hypothesis (see Lovelock, 2000b for a personal account of the history of the model). That is that Gaia looked to many people almost like a religious idea, suggesting that things had to work a certain way to make a planet fit for life. In classic versions of Daisyworld, a cloudless planet is the home of two kinds of plants—conventionally called daisies (Watson and

Lovelock, 1983; Lenton and Lovelock, 2001). One type is dark coloured (and so ground covered by it reflects less light than does bare ground) and the other is light coloured (reflecting more light than bare ground). These daisies are usually referred to as 'black' and 'white', although the key point is the way in which their albedo differs from unvegetated ground. The growth rates of these daisies are assumed to be a parabolic function of temperature. Black daisies absorb more light energy than do white ones and so become warmer. The temperature of the planet Daisyworld is governed by the input of stellar energy and the albedo of the planet, so life and the physical environment are therefore coupled in the model (see Box 7.1 for a summary of the mathematics underling the model).

To understand the behaviour of a simple Daisyworld model, consider a planet that is initially cold but is slowly receiving more stellar energy as its star warms; this is true for Earth, since the sun was fainter in the past (Sagan and Chyba, 1997). The planet reaches a temperature that enables daisies to grow and their seeds to germinate; however, as conditions are still cool the black daisies grow

Box 7.1 An introduction to the mathematics of Daisyworld

The original mathematical description of Daisyworld was given by Watson and Lovelock (1983); for a summary see Wood et al. (2008). Most investigations of Daisyworld have relied on simulations; however, analytical solutions to Daisyworld models are described by Saunders (1994) and Sugimoto (2002). An excellent example of using Daisyworld as a teaching tool, to introduce ideas about planetary systems and feedbacks, is given in chapter 2 of Kump et al. (2010).

The population dynamics of the plants are described using the epidemic model used by Carter and Prince (1981) to model aspects of plant biogeography. Thus the equation for black daisies is:

$$d\alpha_w/dt = \alpha_w\,(x\beta_w - \gamma) \qquad (1)$$

Each daisy population is described as its fractional cover of the planet α (as Daisyworld is a mathematical planet it is assigned an area of 1 for convenience). β is the growth rate per unit time and area while γ is the death rate per unit time. The subscripts 'w' and 'b' (in equations that follow) signify white and black daisies. Studies by Maddock (1991) suggested that Daisyworld was not particularly sensitive to the type of population model used. 'x' is the fraction of the planet's surface not covered by daisies, so that:

$$x = 1 - \alpha_w - \alpha_b \qquad (2)$$

In standard Daisyworld models the plants have an optimum growth temperature of 22.5°C and limits of tolerance to temperatures of 5 and 40°C. The growth rate (β) of each daisy type is a parabolic function of the local temperature (T_i):

$$\beta = \max\left(0.1 - ((22.5 - T_i)/17.5)^2\right) \qquad (3)$$

The effective temperature of the planet (T_e) is given by the following energy balance equation:

$$\sigma(T_e + 273)^4 = SL\,(1 - A_p) \qquad (4)$$

where σ is the Stefan–Blotzmann constant, S is a constant energy flux (so that $S/\sigma = 1.68 \times 10^{10}\ K^4$), and L is a dimensionless measure of the luminosity of Daisyworld's star (273 appears in equation 4 to keep all temperatures in degrees centigrade, this being the freezing point of water on the Kelvin scale). A_p is the albedo of the planet, which is given by:

$$A_p = \alpha_g A_g + \alpha_b A_b + \alpha_w A_w \qquad (5)$$

where A_g is the albedo of bare ground, A_b is the albedo of black daisies, and A_w is the albedo of white daisies—on Daisyworld $A_b < A_g < A_w$. The local temperature of black daisies can be determined from the following equation that relates local temperatures to planetary temperatures:

$$(T_b + 273)^4 = q\,(A_p - A_b) + (T_e + 273)^4 \qquad (6)$$

where $q = 2.06 \times 10^9\ K^4$ and is a measure of insulation between the various regions on the planet's surface. A similar equation can be written for white daisies.

This original version of Daisyworld is zero-dimensional; that is there is no explicit representation of space. More spatially complex versions of Daisyworld also exist; examples include a one-dimensional model with latitudinal variations in stellar radiation (Adams et al., 2003) and a two-dimensional cellular automata model with a curved planetary surface (Ackland et al., 2003).

better than the white ones, because their dark colour allows them to absorb more stellar radiation and so heat up. The black daisies will therefore come to dominate the initial community. As these dark plants spread, they warm the planet, providing a positive feedback which enhances the spread of life on the planet. However, once the planetary temperature gets high enough, the white daisies have a selective advantage over the black daisies. Now the white population expands at the expense of the black, which tends to cool the planet by increasing its albedo—that is reflecting more solar energy back into space. However, as the sun gets hotter, eventually only white daisies are left. As the sun gets even hotter, these plants are finally killed off as the temperature rises beyond the level that allows plants to function.

The result of the daisies' behaviour is a considerable degree of planetary temperature regulation. Over a substantial range of luminosities the daisies act as a negative feedback on planetary temperature; this is illustrated in Figure 7.2 which shows results from a slightly more complex model with three different colours of daisies. The reason for this negative feedback, and the resulting temperature regulation, is that any increase in a daisy of a particular colour tends to force the planetary temperature towards conditions that are less good for the growth of that type of daisy. For example an increase in black daisies will increase planetary temperature, eventually forcing the temperature into a range where black daisies grow less well than the cooler white daisies; this leads to a reduction in black daisies and a decrease in temperature. An important aspect of conventional Daisyworld is that the plant's adaptation (colour) has the same effect on the plant (warming or cooling) as it does on the planet, and obviously this will not always be the case in the real world. However, Daisyworld illustrates the potential for biology to play an important role in regulating conditions on a planet, but for this to happen it requires life to cover much of the planet's surface. If the daisies were restricted to a small fraction of Daisyworld then their effects on the colour of the planet would be negligible and their ability to regulate its temperature would be lost. Here is another potential reason for expecting life on any planet with a persistent ecology to be a planet-wide phenomenon, as it allows life to be involved in regulating processes of planetary significance.

Daisyworlds have attracted a lot of attention and show surprisingly interesting behaviour for such simple models. Much of the work on Daisyworld models originally came from people with a physics or climatology background (see Lovelock, 1992 and Harding and Lovelock, 1996 for important early attempts at more ecological approaches). Wood et al. (2008) provide a detailed review of the many variations on Daisyworld models and their applications. More recently a number of other approaches to modelling the effects of coupling life and the abiotic environment in a somewhat more realistic way have been published—for example they don't make the Daisyworld assumption that what is good for the individual will be necessarily good for the whole planet (e.g. Williams and Lenton, 2007; Dyke and Weaver, 2013; Arthur and Nicholson, 2017). However, as Toby Tyrrell (2013) points out, despite its simplicity 'the original Daisyworld is an extremely satisfying intellectual invention. It is extraordinary that such a simple mechanism can achieve so much'. This simplicity is Daisyworld's great strength (with the necessary downside of a lack of realism) and in this book I am primarily using Daisyworld as a simple conceptual model to illustrate the potential for life to be involved in the regulation of planetary processes if it covers a large enough area of a planet.

7.3 Examples of the role of life in planetary physiology on Earth

Is there anything operating like the black and white daisies on Earth? One possibility is the boreal forest; this covers some 15.8 million km^2, with another 3.3 million km^2 of coniferous forest in temperate mountain areas (Archibold, 1995). Much of the area covered by this great expanse of trees has lying snow, especially over winter. However, these coniferous tree species often shed their snow so that they remain dark coloured, compared to the high albedo of the surrounding snow-covered ground (Figure 7.3). Even where snow survives on the tree canopy, multiple reflections within the canopy scatter the light rather than reflecting it, so the vegetation still has a lower albedo than

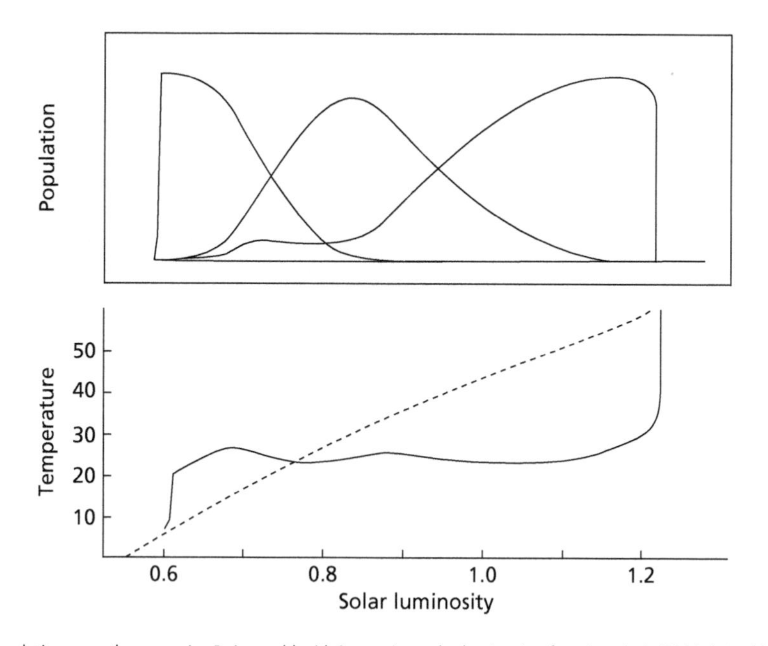

Figure 7.2 Climate evolution on a three-species Daisyworld with increasing solar luminosity, from Lovelock (2000a). In this version a grey daisy species was included as well as the classic black and white daisies. These grey daisies were introduced to attempt to counter the criticism that 'cheats' which didn't have the cost of making pigments could undermine the system. This idea that cheats can destroy otherwise cooperative interactions between different species is a common theoretical problem in evolutionary ecology (Yu, 2001). Lovelock (2000a) added a minor cost to pigment production for black and white daisies. Upper graph shows population sizes of the three daisies (black peaking at lower solar luminosities, then grey, then white). Lower graph shows the temperature (in degrees centigrade) on Daisyworld with plants (solid line) and without any life (dashed line). Clearly, grey daisies did not undermine temperature regulation, although one possible criticism is that all three daisies have seedbanks of infinite lifespan, so however well the cheat did it could not cause the extinction of the other species.

a snow-covered ground surface (Harding and Pomeroy, 1996).

This raises the possibility that the boreal forest may actually warm high-latitude areas, so potentially helping to maintain a climate suitable for trees. An interesting question that follows from these observations is 'What is the net climatic effect of boreal forest in the context of global heating?' We normally think of forests as having a beneficial effect by acting as carbon sinks and so potentially reducing global heating, but the net warming effect of creating a lower albedo at higher latitudes (that is making the land darker in colour) would tend to counteract the 'cooling' effect of carbon sequestration. Modelling suggests that 'in many boreal forest areas, the positive forcing induced by decreased albedo can offset the negative forcing that is expected from carbon sequestration' (Betts, 2000).

Clearly the boreal forest could potentially act like a black daisy. What about white ones? One possible example of a white daisy is some species of marine algae. In this interesting example the 'white daisies' are not the algae themselves but an indirect by-product, namely clouds. Dimethylsulphoniopropionate (DMSP) is found at high concentrations (100–400 mmol/l) in the cells of many marine algae, including coccolithophorids and dinoflagellates. DMSP has a range of possible functions in algal cells including osmoprotection and as an antioxidant (Simó, 2001; Sunda et al., 2002). DMSP breaks down to dimethylsulphide (DMS), which is the most common form of volatile sulphur in the oceans and a major source of cloud condensation nuclei—that is the small particles around which water droplets can form to make clouds. This gives the possibility of a Daisyworld-like mechanism, often referred to as the CLAW hypothesis after the initials of the authors of the original *Nature* paper suggesting this mechanism (Charlson, Lovelock, Andreae, and Warren, 1987). I once asked James Lovelock if they had spotted that their initials would make a word

Figure 7.3 On a small scale this photograph, taken in the Swiss Alps, illustrates Richard Betts' (2000) point about the boreal forest. The snow shedding shape of these coniferous trees makes them black 'daisies' in a white winter landscape. The dark trees absorb heat, while the white snow reflects it. Another way to understand the potential importance of albedo is to think of two flat rocks left out in the sun, one light coloured and the other black—the black rock will heat up more than the light-coloured one in the sun and may even become uncomfortably hot to the touch. The importance of albedo doesn't just apply to the extreme case of dark forest and white snow For example while evergreen coniferous forest typically has an albedo of 8–12% (broadleaved deciduous forest is 14–18%), typical values for arable crops in temperate locations are 20–25%—so such crops (and grasslands) reflect more strongly than forest (Bateman et al., 2023).

and if this had affected the order they listed the authors (a ruse that has occasionally been used to raise the profile of a research paper), but he said that the four authors had just been listed in order of their contribution to the work. The CLAW paper is now a citation classic, with over 5,000 citations in *Google Scholar* (as of early 2023).

The basic idea was that warmer conditions potentially favour algal growth, so leading to more DMS and thus more cloud reflecting solar radiation and therefore cooler conditions which then lead to less DMS production. This is a negative feedback system. Just as when Daisyworld gets warmer the white daisies expand in their coverage and so cool the planet, on Earth as algae expand they make more of a type of cloud that cools the planet (the role of clouds can be complex; for example some very high-altitude clouds warm, rather than cool, the planet). Over the last 35 years the explosion in research in this area, much of it triggered by the CLAW paper, has developed this relatively simple idea. For example, it is now possible that under glacial conditions the mechanism may operate as a positive feedback loop, making the planet even colder. This could happen because many marine algae are limited by nutrient levels in the ocean (especially iron). In the arid conditions of glacials much more dust may blow into the oceans, adding nutrients and so increasing algal growth and DMS production, and therefore increasing cloud cover (reviewed by Welsh, 2000). An additional complication is that colder conditions can cause thermal stratification of the oceans, isolating the surface layers from more nutrient-rich waters below. More work is needed to sort out the complexities of this system—indeed there is still no general consensus on the validity of the details of the CLAW hypothesis (Jackson and Gabric, 2022). Despite this, however one interprets the details, it is clear that the CLAW hypothesis gave rise to a whole new area of interdisciplinary research on the links between marine algae and clouds (Nunes-Neto et al., 2009).

Irrespective of the overall sign of this feedback loop, the DMS system illustrates microbial life having a large, and potentially regulating, effect on a planet's climate. In addition, this DMS plays a major part in the sulphur cycle by transferring large quantities of this element from the oceans to the land; indeed it was this postulated role in cycling sulphur that originally caused James Lovelock to look for DMS production by marine algae in the early 1970s (Lovelock, 2000b; Simó, 2001). Through the production of DMSP, the physiology of algal cells plays an important role in the physiology of the planet, by both cycling sulphur and affecting the climate.

The bog mosses *Sphagnum* spp. also have much in common with the plants of the planet Daisyworld. These mosses are characteristic plants of many peat bogs and thrive in wet conditions. *Sphagnum* can hold up to 20 times its own weight in water with the aid of special empty 'hyaline' cells that allow it to act like a sponge. As such the moss makes its own local environment wetter, and so more suitable for itself, while also making the whole bog ecosystem wetter (Porley and Hodgetts, 2005). Just as the model 'daisies' have the same local and global effects (warming or cooling) *Sphagnum* has a similar local and ecosystem-wide effect with water content (and also with pH, increasing the acidity both locally around its cells and more widely across the bog system). *Sphagnum* is a supreme ecological engineer, as Porley and Hodgetts (2005, p 361) have written—'*Sphagnum* does not just live in the bog: it *is* the bog'—helping to control the moisture content and acidity of the whole system, while often physically constructing the bog from its own peat-forming remains (Figure 7.4). However, despite these putative examples of black and white 'daisies' in the Earth system, it seems too much of a stretch to think that most global regulation could come from fortuitous cases where the effects of the organism's trait on the individual is identical to its wider environmental effects. This is a problem returned to in more detail in chapter 12.

Figure 7.4 '*Sphagnum* does not just live in the bog: it *is* the bog'. The top 13 cm of *Sphagnum* peat sampled from a bog in the Swiss Pre-Alps in winter (this sample formed part of the study by Lamentowicz et al., 2013). The top 3 cm are frozen, but below this the fibrous nature of the peat—comprising partly decomposed *Sphagnum* moss—can be clearly seen.

Life has been implicated in the regulation of many other aspects of the Earth system; examples include carbon dioxide and oxygen levels in the atmosphere (for O_2 see chapter 8 and CO_2 chapter 9). Another important example is the nutrient balance of Earth's oceans. Classic studies by Alfred Redfield in the 1930s identified a clear empirical relationship between the physiology and biochemistry of plankton and the physiology and chemistry of the oceans (Redfield, 1958; Falkowski and Davis, 2004). Redfield pointed out that the nitrogen:phosphorus ratio of many plankton was approximately 16:1 and that the N:P ratio of seawater was also very similar (these are now often called Redfield ratios). This strongly suggested a major role for plankton in regulating ocean chemistry through a range of processes including the acquisition of nutrients by plankton, the formation of new planktonic biomass, and the remineralization of this biomass by bacteria (for reviews see Falkowski and Davis, 2004 and/or Sherratt and Wilkinson, 2009).

Our lack of a detailed understanding of these systems is highlighted by the relatively recent discovery that the rate of N-fixation in the open ocean appears to be much higher than previously thought (Capone et al., 2005). Prior to this work it had been thought that most nitrogen in the ocean's euphotic zone (the well-lit upper layer of the ocean where photosynthesis happens) came from the mixing of nitrate from deeper waters rather than from planktonic N-fixation. Indeed, it seems that N-fixation by photosynthetic bacteria plays an important role in the negative feedback loops that help maintain the ocean at Redfield ratios. If the rate of phosphorous input (via rivers from weathering on land) rises then the N-fixation allows plankton numbers to expand to use this increased phosphate, thus bringing the N:P ratio back to around the Redfield values (Tyrrell, 1999; Lenton and Watson, 2000a, 2011). The Redfield ratios are a particularly spectacular example of organismal physiology playing a major role in the physiology of the whole planet, but to have this effect requires a high biomass of plankton. Indeed these ratios appear to be an average value that emerges from the different physiologies or a range of phytoplankton growth strategies rather than a global optimum value for all plankton (Arrigo, 2005); this makes their persistence even more remarkable.

7.4 The importance of biomass: an illustration from Earth's past

Just over 65 million years ago at the end of the Cretaceous, a dramatic event apparently triggered one of the five great mass extinction events known from the fossil record. These Phanerozoic mass extinctions (there were no doubt others during the earlier Proterozoic and Archean) are of more than historical interest as we may currently be causing a 'sixth extinction' through our own actions (Leakey and Lewin, 1995). There is currently a high level of agreement that a large meteorite impact was associated with the end Cretaceous extinction, although volcanism, especially in the Deccan Traps in India, may also have played a role (Cleland, 2020). It also seems plausible that climate change may have already been stressing an ecology adapted to a hothouse climate prior to the meteorite's impact (Renne et al., 2013; Condamine et al., 2021).

The probable impact site has been identified in the Yucatan peninsula in Mexico; in this area the carbonate rocks were anhydrite ($CaSO_4$) rich, and an impact here would have caused large amounts of sulphur dioxide (SO_2) to enter the atmosphere on impact. This increase in atmospheric SO_2 would have been detrimental to many life forms as it would have caused extensive acid rain; however, it would also have acted as an ultraviolet light shield, helping to compensate for damage done to the ozone layer by the impact (Pope et al., 1998; Cockell and Blaustein, 2000). A combination of palaeobotanical, isotope, and modelling work has suggested that the terrestrial biomass recovered extremely quickly (possibly in less than 100 years) from this global catastrophe, while the marine system took some 3 million years to return to its pre-impact productivity (Beerling et al., 2001a; Lomax et al., 2001). So biomass recovered on a much faster scale, at least on land, than was required for the evolution of replacement species, so *productivity recovered much faster than biodiversity*. This conclusion is highly relevant to the discussions of the role of biodiversity in academic ecology, which has tended to focus on much shorter timescales. For example high biomass forests have returned to much of eastern North America since the decline of farming in the nineteenth and early twentieth centuries (Foster, 1992) and an entirely human-made forest has formed on the previously treeless Ascension Island over a similar timescale (Wilkinson, 2004a). Given a suitable climate and soils, high biomass vegetation can easily develop in a timescale of 100 years or less if ecologically appropriate plant species are available.

These palaeobotanical data are of interest in the context of Grime's (1998) suggestions on the functioning of modern plant communities. He argued that 'the extent to which a plant species affects ecosystem function is likely to be closely predictable from its contribution to the total biomass'—an idea he called the 'mass ratio' hypothesis. In other words, it is the total biomass (normally made up of a very limited subset of species) that governs most ecosystem services; the majority of rare species contribute very little. For example, Smith et al. (2020) found that in experimental studies of tall grass prairie plant communities, mass ratio effects, rather than the effects related to species richness, were key in understanding ecosystem function. And Listner et al. (2023) recently demonstrated an important role for the mass ratio effect—and a rather limited role for species richness—in biomass production in meadows in the Czech Republic. The end Cretaceous extinction event is consistent with this in that a vegetation that had lost many of its species apparently returned to a large biomass, with its associated role in carbon cycling, etc., possibly on a timescale of decades. As such, it was biomass (not species richness) that was crucial to the recovery. However, it is possible that the more species rich the vegetation was before impact, the greater the chance of it containing species suitable for post-impact conditions, and this may be the most important role for biodiversity when viewed at a planetary scale (see section 4.5).

7.5 Biomass and Gaia

The arguments I have developed above make a plausible case that life on a planet will be widespread and therefore have a substantial biomass and that because of this biomass the products of these organisms' physiology will play a large and potentially regulating role in the physiology of the planet (this chapter). It is clear that there is the potential for systems involving life to have a self-regulating character, but what about abiotic systems? The natural nuclear reactors that existed 2 billion years ago at what is now Oklo in Gabon, West Africa are a good illustration of a self-regulating natural system that probably didn't have any significant biological involvement and hence shows that life is not necessary for regulation. These geological nuclear reactors comprised uranium ore in sandstone rocks overlying impermeable granite and were probably naturally moderated by water; if the reactors ran too fast this water turned to steam, so reducing its ability to slow the neutrons and prevent them being absorbed by naturally occurring ^{238}U (Maynard Smith and Szathmáry, 1995; Meshik, 2005). This controlled the rate of the nuclear reactions in a manner analogous to that used in nuclear power plants. In this case there is a possibility that bacteria, some of which can sequester uranium, may have been involved in the initial formation of these ore deposits (James Lovelock, pers. comm.); however, even if this was so, the likelihood is that the regulation of these reactors was largely abiotic. Since regulation does not require biology, a key question is 'Are there any theoretical reasons to expect systems involving life to be more likely to be self-regulating that solely abiotic systems?'

In the context of Earth, a plausible answer to this question is that Earth systems do appear to have regulated conditions within a life-friendly range over geological time. However, this is quite possibly accidental, with no reason to expect the same is likely to happen on any other planet with life. Several people raised this possibility in the context of Gaia theory around the end of the 1990s (Lenton, 1998; Watson, 1999; Wilkinson, 1999); however, it is Andy Watson who has developed these arguments in most detail (Watson, 1999, 2004). The basic idea is easily explained and has much in common with the anthropic principle in astronomy, where the presence of astronomers clearly implies aspects of the nature of the universe (Carr and Rees, 1979; Hoyle and Wickramasinghe, 1999). As I have previously written: 'Any planet which is home to organisms as complex as James Lovelock and Lynn Margulis must have had a long period of time during which conditions were always suitable for life, and thus must give the impression of regulation for life-friendly conditions even if the persistence of life was purely a matter of chance' (Wilkinson, 2004b, p 72).

As long as these arguments are based on a single example—Earth—it is impossible to rule out an explanation based entirely on chance. However, such an explanation does seem rather unlikely (but see Tyrrell, 2013). If life evolves very rarely—or just once—on a planetary surface then it is remarkably unlikely that chance would provide a self-regulating system. If one assembles biospheres in such a random way then it would seem likely that there would be many more ways of producing non-regulating systems than regulating ones. If we assume that life arises on planets reasonably frequently then there is the possibility of a selection process, so that only planets that develop self-regulating systems—even if this is 'by chance'—are likely to survive for long as a living planet. However, it seems likely that the full explanation involves special properties of planetary systems involving life.

The most important, and provocative, part of Gaia theory is that a planet's environment is regulated 'at a habitable state for whatever is the current biota' (Lovelock, 2003, p 769). Imagine a planet where the climate and chemistry are currently in a suitable state for its prevailing life forms—an easy task as we live on such a planet. If the activities of life were tending to force this system towards an uninhabitable state, for example through the production of some metabolic by-product, then as the planet approached this uninhabitable state these organisms would increasingly struggle and so the forcing would be reduced (an alternative is that natural selection may produce something that consumes the by-product; for example see section 3.1 on lignin). These inbuilt regulating mechanisms, which by definition work to keep the system in life-friendly conditions, are not present with purely abiotic forcings. With the evolution of a significant biomass of more complex organisms the range over which this regulation happens will tend to decrease; for example the upper temperature limit for prokaryotes is higher than for single-celled eukaryotes, which are less temperature sensitive than vascular plants (Table 1.1). This is the significance of the phrase 'whatever the current biota', in Lovelock's definition of Gaia cited above. Just as Grime (1998) argues that the high biomass of dominant species largely controls the behaviour of a plant community—the mass ratio hypothesis—so in a similar manner the dominant life forms (by biomass, rather than by number of species or individuals) may control the properties of a planet's biosphere—at least until the rise of an intelligent species with an industrial economy that can have a planetary influence out of proportion to its biomass.

This importance of biomass connects with the idea of Gaian bottlenecks, briefly introduced in chapter 6. This starts from the observation that although the basic requirements for life seem common in the universe, we have no evidence to suggest the universe is teaming with life. The standard answer to this is to assume that the emergence of life is a very unlikely event. The Gaian bottleneck approach suggests that there is a second possibility (not necessarily mutually exclusive) that if 'life emerges on a planet, it only rarely evolves quickly enough to regulate greenhouse gasses and albedo' (Chopra and Lineweaver, 2016, p 7). To do this, life would need a biomass sufficient for the organisms' physiology to have planet-wide effects. Recent modelling of virtual planets hosting microbial biospheres—the ExoGaia model (Nicholson et al., 2018a)—provided clear illustrations of Gaian bottlenecks, with life on some planets becoming extinct before it had managed to establish the regulation needed for long-term habitability.

These Gaian mechanisms can be thought of as loading the dice in favour of the long-term survival of life once it has started on a planet. In the 'just chance' anthropic view the survival of life is like rolling a 6 with a fair dice (although the probability of 'success' is likely to be much less than 0.17), while Gaian mechanisms load the dice to make a 6 much more likely—although probably not inevitable. Therefore key questions of a Gaian view of planetary ecology are 'What are the mechanisms?' and 'How strongly do they load the dice?'

7.6 Overview

All organisms modify their environment to some extent, either as an evolved adaptation (e.g. beaver dams) or through by-products of their activity (e.g. oxygen release by photosynthesis). Daisyworld models illustrate the theoretical possibility that these modifications could lead to planetary regulation for life-friendly conditions, but also illustrate the need for substantial biomass for this to occur. Examples of possible regulatory processes on Earth include the production of DMSP by many marine phytoplankton and the stability of the N:P ratio of seawater. Crucially for life to have these effects it must have a considerable biomass, so that its products can significantly influence a planetary system. In academic ecology the recent concentration on biodiversity may have tended to obscure the fundamental importance of biomass—which will often be dominated by a limited number of species. This 'mass ratio' effect is important at both the ecological community and planetary scale. Once biomass is sufficient for the merging of organismal and ecological physiology then the planet will tend to show some level of regulation for life-friendly conditions; as such this merging—and its associated high biomass—has a positive Gaian effect. The magnitude of this expected effect on any planet is currently an open question. It could range from a slight increase in the probability of life surviving to much tighter regulation reminiscent of homeostasis in an organism. The size of this effect is currently a key unknown in both Earth-based environmental science and astrobiology.

CHAPTER 8

Photosynthesis

8.1 Quantification and mysticism in the seventeenth century

Imagine trying to understand how a plant grows if you had not been taught any biochemistry. Animals clearly eat to gain resources, but plants? An obvious guess is that they are eating soil, but a famous seventeenth-century experiment shows this to be wrong. The Flemish chemist and medic Jan Baptista van Helmont (1579–1644) planted a willow sapling in a pot and found that after five years of watering the tree had increased in weight by 74.1 kg (and that was excluding four years' worth of fallen leaves); however, the soil had lost negligible weight. His involvement in this long-term experiment was perhaps encouraged by the fact that he had been forbidden to travel by the Inquisition (Ball, 2005b). Van Helmont's interpretation was that the plant's increase in weight must have come from the water; the correct answer that it had largely come from the 'air' would have seemed incredible. The construction of a tree from elemental water made sense to Van Helmont who was a vitalist and saw life in everything, even the rocks. Indeed, he regarded water as a primary element which could form plants and much else. While a proponent of the experimental method he was also very at home with a magical view of the universe—part 'scientist' and part mystic (Porter, 1997; Henry, 2002). This was a world view so odd to modern eyes that the historian Herbert Butterfield (1957, p 129) thought that some of the strangest ideas of other scholars in Van Helmont's time seemed 'rationalistic and modern in comparison'. While this magical approach seems odd from a twenty-first-century perspective it was not an unusual world view for the natural philosophers of the seventeenth century, an obvious example being Isaac Newton's interest in alchemy (Dobbs, 1982).

The ability of photosynthetic organisms to use such common resources as carbon dioxide, water, and light to make carbohydrates has big implications for Earth's ecology, and raises interesting questions about the likelihood of photosynthesis evolving on any planet with Earth-type life. The release of oxygen as a waste product of oxygenic photosynthesis has been a key factor in Earth's ecological history and the evolution of its atmosphere.

8.2 The diversity of photosynthesis on Earth

When most ecologists think of photosynthesis they tend to picture a green plant (Figure 8.1), such as van Helmont's willow sapling; however, on a global scale this is rather misleading. On Earth about half of the primary production is produced by marine microbes (Falkowski, 2002; Arrigo, 2005). Our knowledge of these has increased tremendously over the last few decades, but it has not yet attained much prominence in the education of most ecologists.

One of the key figures in the ecology of marine plankton during the first part of the twentieth century was Alister Hardy. In his classic account of the natural history of plankton (Hardy, 1956) he gave a snapshot of the state of knowledge at the time. In describing photosynthetic phytoplankton he mainly wrote about the larger taxa such as diatoms, but briefly mentions that there are 'vast numbers' of smaller eukaryotes which pass 'through the mesh of the finest net we can use' (Hardy, 1956, p 49). Of prokaryotes he wrote: 'smaller still are the

The Fundamental Processes in Ecology. Second Edition. David M. Wilkinson, Oxford University Press.
 DOI: 10.1093/oso/9780192884640.003.0008

Figure 8.1 Plant diversity is a small part of the total diversity of photosynthetic organisms. Shown is alpine pasqueflower *Pulsatilla alpina*, growing at around 2,000 m near the summit of Rochers de Naye in Switzerland. It's a member of the Ranunculaceae (buttercup family) which contains around 2,346 species. The largest plant family is the Orchidaceae with around 26,470 species. There are an estimated 321,000 plant species globally, with 451 vascular plant families, according to the Angiosperm Phylogeny Group (APG) IV classification of 2016 (Christenhusz et al., 2018). To give an indication of how our knowledge of plants has changed over the last 200 years, at the start of the nineteenth century the botanist Robert Brown—today probably most widely known for giving his name to Brownian motion—estimated that around 33,000 plant species were known to science (Brown, 1814). However, as described in the main text, much of global photosynthesis (and oxygen production) is carried out by a diverse range of microorganisms rather than true plants, and we have much less idea of the total diversity of photosynthetic microbes. Indeed there is considerable uncertainty over how to define species in prokaryotes such as cyanobacteria, or even if the concept of species makes sense for an organism with substantial amounts of lateral gene transfer (O'Malley, 2014).

bacteria ... at present very little is known about their occurrence in the plankton' (Hardy, 1956, p 50). Since Hardy's book it has become apparent that very small phytoplankton are enormously important in both marine and freshwater habitats (Figure 8.2)—the marine taxa being the most important in the Earth system because of the vastly greater extent of the oceans. These microbes are often referred to as the picophytoplankton and can be defined as photosynthetic O_2-evolvers that can pass through 2-μm-diameter pores (Raven, 1998), much smaller than the 'finest bolting cloth' used by millers for sieving flour which Hardy (1956) recommended for making plankton nets.

Some of the most important members of the picophytoplankton are the cyanobacteria, one of the key taxa being *Prochlorococcus* spp. Although not described until 1988, this genus dominates primary production in the tropical and subtropical oceans and must be a good candidate for the title of the commonest organism on Earth (Fuhrman, 2003; Sullivan et al., 2003). It has a diameter of about 0.6 μm and the smallest genome of any photosynthetic organism yet studied (1,716 genes in the ecotype adapted to high light levels: Rocap et al., 2003). Along with the slightly larger (0.9 μm) *Synechococcus* spp. these plankton are major contributors to primary production. In the context of my

Figure 8.2 Collecting plankton at Bantou Reservoir, near Xiamen, southern China. To the left a plankton net is just being pulled up after being used to sample the water column. Behind can be seen the outflow pipe attached to a pump that can be used to sample water from known depths; in the foreground is a probe for recording aspects of the abiotic environment—such as temperature, pH, and dissolved oxygen content of the water. This reservoir forms part of long-term studies of aquatic microbial ecology by the Chinese Academy of Sciences Institute of Urban Environment in Ximen (e.g. Liu et al., 2019).

discussion of the potential role of parasites in population regulation and evolution (sections 3.4 and 3.5) it is interesting to note that these picophytoplankton are the hosts for a range of viruses; indeed viruses appear to be an important and previously understudied aspect of marine ecology (Suttle, 2005; Biggs et al., 2021). The extent and importance of marine viruses became apparent during the 1990s, and research in this area rapidly expanded during the first two decades of the twenty-first century (Gilbert and Mitra, 2022).

On land, we have long described the main biomes based on a mix of climate and the dominant types of photosynthetic plants (often trees), such as tropical forests, tropical savannas, Mediterranean ecosystems, and temperate forests (e.g. Schimper, 1903; Archibold, 1995). However, these biomes also contain photosynthetic microbes and in some systems with few true plants they can be very important. For example Hodkinson et al. (2004) studied the invertebrates along a chronosequence created by a retreating glacier on the Arctic island of Svalbard. When I examined sediments from their sites under the microscope it was obvious that there were large numbers of cyanobacteria, even in sediments that had only been exposed by the retreating glacier for a couple of years. Indeed these microbes dominate many terrestrial and freshwater polar habitats and play an important role in the carbon and nitrogen cycles at these sites (Vincent, 2000). They have also been considered important in desert biocrusts (Figure 1.2), but until very recently have been little studied in most terrestrial habitats. However, soil photoautotrophic microbes—both cyanobacteria and micro-eukaryotes—appear to be much more important than previously assumed. For example, Jassey et al. (2022) estimate that soil algae are responsible for around 6% of net primary productivity of terrestrial vegetated landscapes. That equates to just under one-third of global human carbon emissions. While much of this carbon may be rapidly cycled through the algae, and so not sequestered long term, it may significantly contribute to soil organic carbon (which is an important carbon sink). This is an area of ecology that clearly needs more research attention.

So far in this chapter I have attempted to give an overview of the taxonomic and structural diversity of photosynthetic organisms, from bacteria to large trees. However, there is also important biochemical diversity; for example the range of types of carotenoids is much greater when we consider all photosynthetic organisms rather than just green plants (Box 8.1).

Box 8.1 Carotenoids

Carotenoids are yellow, orange, or red pigments found in all photosynthetic organisms, with a much greater diversity being exhibited by the photosynthetic prokaryotes than plants. They perform several useful functions in true plants, mainly in the folding and maintenance of various photosynthetic proteins and in photoprotective mechanisms that allow excess light energy absorbed by chlorophylls to be dissipated as heat and to deactivate potentially damaging reactive oxygen 'species' (Förster and Pogson, 2004). In addition, they can absorb light and so potentially act as accessory light-harvesting pigments; but this function is probably not very important in true plants, since they absorb light in the same wavelengths as the chlorophylls. However, in some photosynthetic bacteria they absorb light over wavelengths different from the bacteriochlorophylls (Madigan et al., 2012). This raises interesting questions about their evolution; while most plant scientists are used to thinking of them as primarily photoprotective, could they have first evolved wholly or partly as photosynthetic pigments?

Carotenoids are most obvious to people living in temperate zones during autumn (fall), when they are one of the main sources of autumn leaf colour (Figure 5.1); indeed the xanthophylls (one of the main classes of carotenoids) were named after their contribution to autumn colour—*xanthos* meaning yellow and *phyll* meaning leaf (Förster and Pogson, 2004).

There is also diversity in the basic chemistry of photosynthesis. The summary equation for oxygenic photosynthesis is well known and will have been familiar from school science classes to most readers of this book:

$$CO_2 + 2H_2O \rightarrow \text{light} \rightarrow (CH_2O) + O_2 + H_2O$$

This involves splitting water using energy from sunlight to obtain electrons and protons (the oxygen is given off as a waste product), the electrons being further boosted in energy before being transferred to carbon dioxide where they provide the energy to make sugars. Lenton and Watson (2011), whose account of the basics of oxygenic photosynthesis I have just paraphrased, point out that the use of 'water as a source of electrons was an evolutionary "stroke of genius" by the first cyanobacterium, because water was already an absolute requirement for life' so anywhere an organism lived must have available water.

Most green plants achieve this by so called C_3 photosynthesis, where CO_2 is fixed into the three-carbon molecule 3-phosphoglyceric acid (commonly shortened to PGA). The most common variant on this is C_4 photosynthesis, where the initial products are fixed as four carbon organic acids which are moved to where they are needed for photosynthesis within the cell, reducing problems with photorespiration—that is the competition between O_2 and CO_2 for sites on the enzyme rubisco (ribulose-1, 5 bisphosphate carboxylase-oxygenase) (Mooney and Ehleringer, 1997). The discovery of C_4 photosynthesis in the mid-1960s stimulated much biochemical research on photosynthetic pathways and on the ecological importance of these pathways (Hibberd and Furbank, 2016). The other common variant on plant photosynthetic pathways is crassulacean acid metabolism (CAM), which had been discovered somewhat earlier; here the separate carboxylation takes place in the same location within the leaf but separated in time (Figure 8.3).

This biochemical variation is, however, within just one type of photosynthesis, namely oxygenic photosynthesis. However, some microbes (such as green sulphur bacteria) carry out anoxygenic photosynthesis, where the source of reductant is the hydrogen in H_2S rather than the H_2O familiar from oxygenic photosynthesis (Fenchel and Finlay, 1995). This can be summarized as:

$$CO_2 + 2H_2S \rightarrow \text{light} \rightarrow (CH_2O) + 2S + H_2O$$

Note that oxygen is not a by-product of this process, and this has considerable implications for planetary ecologies.

8.3 Photosynthesis and the Earth system

It has been clear since the early 1970s that the chloroplasts in plant cells were originally derived from prokaryotes (Margulis, 1971; Bonen and Doolittle, 1976; Quammen, 2018); therefore questions about

Figure 8.3 CAM plants comprise approximately 7% of terrestrial plant species and are distributed across some 33 families (Cushman, 2004). The Joshua tree *Yucca brevifolia* (photographed in Mojave Desert, California, USA) is one of the largest plant species to use the CAM variant of oxygenic photosynthesis; this allows it to open its stomata at night to fix CO_2 that is then stored for use in photosynthesis during the day—when light is available. Night-time stomatal opening allows the plant to reduce water loss associated with acquiring CO_2; indeed it can provide CAM plants with a five- to ten-fold increase in water use efficiency relative to C_4 or C_3 plants in environments with strong daily temperature fluctuations (Crawford, 1989; Cushman, 2004). The fact that some plants can fix CO_2 during the night to be used the next day has been known since the late nineteenth century (Warming, 1909).

the early history of photosynthesis on Earth are really questions about the origin of photosynthesis in bacteria. There has been a general assumption by many researchers that anoxygenic photosynthesis predates oxygenic photosynthesis (Fenchel and Finley, 1995), a conclusion supported by more recent molecular analysis (Xiong et al., 2000).

The evolution of anoxygenic photosynthesis on a planet is ecologically important, as it provides a source of energy for organisms that does not just rely on pre-existing organic chemicals. Circumstantial evidence for anoxygenic photosynthesis has been described from 3,416-million-year-old rock in South Africa (Tice and Lowe, 2004). However, the rise of oxygenic photosynthesis is arguably an even more important major transition in the history of Earth's environment, as it led to a greatly increased supply of free energy for life (Lenton et al., 2004; Lenton and Watson, 2011). Aerobic food chains can potentially support many more trophic levels than anaerobic ones (Figure 8.4). Since size is a major factor in structuring many food chains, longer food chains make possible the abundance of macroscopic animals that feature so prominently in many ecology textbooks. In a detailed study of published food webs, Cohen et al. (1993) found that around 90% of the feeding links between animal species with known sizes had a larger animal consuming a smaller prey. Obvious exceptions to this general rule are most parasites. In general, multicellular predation may require an oxygen-rich atmosphere for its evolution, for both physiological and ecological reasons outlined above. These arguments suggest that oxygenic photosynthesis may be intimately connected with a planet's biomass, and hence the likelihood of the merging of organismal and ecological physiology (Chapter 7). Anoxygenic photosynthesis will increase the system's energy source, so potentially increasing biomass. However, oxygenic photosynthesis—if it leads to an oxygen-rich environment—sets the stage for a far larger planetary biomass.

8.4 Oxygen and the Earth system

Currently our atmosphere is 21% oxygen (Canfield, 2014). There are two main possibilities for the source of oxygen in Earth's atmosphere: the

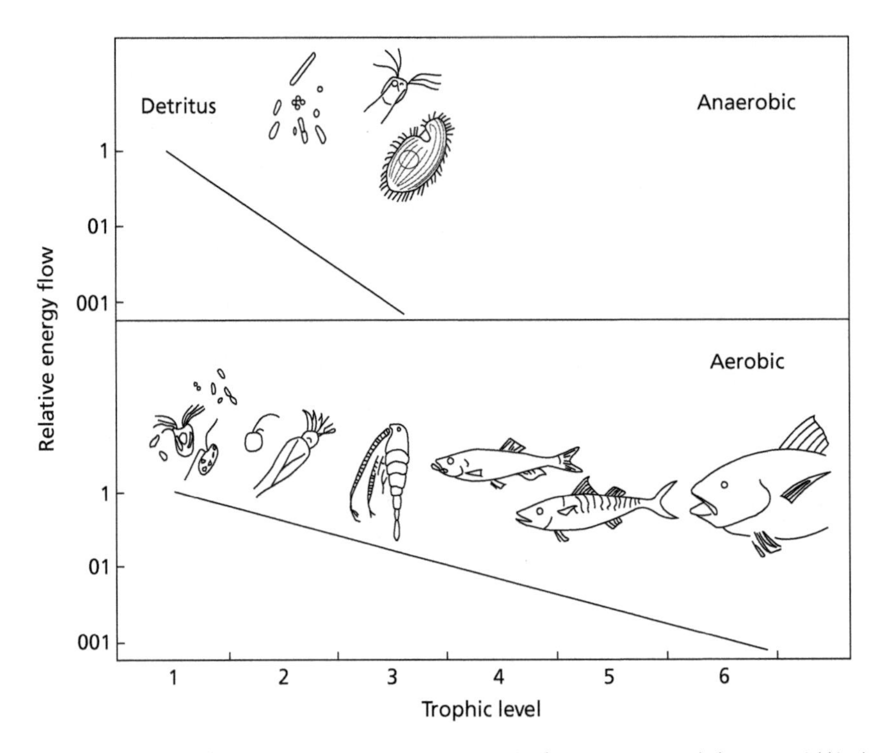

Figure 8.4 Oxygen should allow longer food chains on a planet. In extant anaerobic fermentation on Earth the energy yield is about 25% that of aerobes. Assuming a growth efficiency of 10% for anaerobic food chains and 40% for aerobic ones, a similar amount of energy will support a much longer food chain in an aerobic system. In the example shown here the anaerobic food chain stops at protists, while the aerobic one continues to eventually support large fish. (From Fenchel and Finlay, 1995.)

abiological photo-dissociation of water molecules or oxygenic photosynthesis (Catling and Claire, 2005). The mechanism for the abiological photo-dissociation of water is simple: ultraviolet light is energetic enough to break apart water molecules into hydrogen and oxygen and the hydrogen is light enough to escape Earth's atmosphere into space, leaving spare oxygen behind. However, by the middle of the twentieth century it was apparent that this process couldn't explain the observed levels of oxygen in Earth's atmosphere. One problem is that ozone (O_3) is formed as the levels of diatomic oxygen (O_2) build up in the atmosphere, and this ozone shields water molecules from incoming ultraviolet light, so reducing oxygen formation (Berkner and Marshall, 1972). This ozone 'problem' means that photosynthesis must be the main source of oxygen in the atmosphere. As James Lovelock pointed out in the 1960s, oxygen in Earth's atmosphere is a tell-tale indicator of life on our planet, especially as it is coexisting with methane, which is not a chemically stable situation (Lovelock, 1965; Lovelock and Margulis, 1974). As far as we are aware, photosynthesis is the only process that causes the photo-dissociation of water using visible light—as opposed to more energetic ultraviolet light (Sagan et al., 1993).

Consider this pairing of the summary equations for oxygenic photosynthesis and aerobic respiration:

$$CO_2 + 2H_2O \rightarrow (CH_2O) + O_2 + H_2O$$

$$(CH_2O) + O_2 \rightarrow CO_2 + H_2O$$

Biochemically, aerobic respiration is the reverse of photosynthesis. Think about a dead plant (or cyanobacterium); to a large extent its structure has been made from inorganic carbon fixed by photosynthesis. As the dead plant is consumed by detritivores (or the live plant by herbivores) the organic carbon is converted back into inorganic carbon (CO_2) and re-enters the atmosphere. Therefore when considering the amount of atmospheric oxygen the key factor is the amount of biological material buried in sediments. This is not respired and so there is some oxygen left over from its formation by photosynthesis (see Holmes, 1944 for an early textbook account; also Lenton, 2001; Berner, 2004; and Canfield, 2014 for more recent discussions). Indeed this is why marine phytoplankton are so important in contributing oxygen to the atmosphere as a small fraction of their biomass is buried in marine sediments, while on land most biomass is respired, so using up the oxygen it produced during its lifetime (Kasting and Siefert, 2002). The claims often made by less scientifically informed environmentalists that the rainforests are the 'lungs of the world' because of their claimed oxygen production are incorrect because most of the luxuriant plant growth is respired on its death unless washed down rivers into marine sediments or added to peat deposits. However, there are many other crucial reasons to conserve tropical forests, which often have very high biodiversity and sequester substantial amounts of carbon (see chapter 9).

The date of the first oxygenic photosynthesis and the timing of the rise in oxygen in Earth's atmosphere are the subject of considerable dispute (Nisbet and Fowler, 2003; Lenton and Watson, 2011). While the rise in oxygen cannot have happened before the evolution of photosynthesis, it is likely that it may not have happened until well after the first photosynthetic oxygen production. If the first free oxygen that was produced reacted with chemicals on early Earth (such as reduced ferrous iron) this would have delayed the rise in atmospheric oxygen. As described in section 1.4 many structures classically interpreted as very early microfossils have the appearance of cyanobacteria, but there are now considerable doubts about their biological origin. If these 'fossils' are indeed cyanobacteria, this raises the possibility of an oxygen-rich atmosphere 3.5 billion years ago. One problem with this timing comes from the faint young sun paradox, which points out that the sun was much fainter in the past. To prevent the freezing of all the water on early Earth, some kind of greenhouse effect is usually suggested. While Sagan and Chyba (1997) suggested an ammonia greenhouse, most workers currently assume that methane and carbon dioxide would have been the main greenhouse gases (Nisbet and Fowler, 2003). Much of this methane could have been of biological origin (Box 8.2); clearly a methane greenhouse would be incompatible with high levels of oxygen because the two readily react, following the equation:

$$CH_4 + 2O_2 \rightarrow CO_2 + 2H_2O$$

Box 8.2 Methane

Currently on Earth there are two main types of microbial methanogenesis (Fenchel et al., 1998). The first are microbes which break down acetate according to:

$$CH_3COOH \rightarrow CO_2 + CH_4$$

Examples include *Methanosarcina* and *Methanosaeta*. The second are methanogens (e.g. *Methanobacterium* and *Methanococcus*) which use hydrogen as an electron donor, so that:

$$4H_2 + CO_2 \rightarrow CH_4 + 2H_2O$$

The evolution of methanogens is usually assumed to predate photosynthesis, a conclusion with some molecular support (Gribaldo and Philippe, 2002). One possibility is that life on early Earth starts in rare hot habitats (e.g. hydrothermal vents). Over time microbial methane production would create a strong greenhouse effect since methane is an important greenhouse gas—approximately 20 times more potent than CO_2 (Fenchel et al., 1998). This would 'improve' the climate from the perspective of Earth-type life, allowing it to spread more widely on the planet (Nisbet and Fowler, 2003), because the faint young sun paradox suggests that conditions on early Earth may have been cold. If this scenario turns out to be correct, then methane played a crucial role in making early Earth suitable for a planet-wide ecology. This view of Archean methane is consistent with both Gaia theory (Lovelock, 1979, 2003) and the ideas of niche construction of Odling-Smee et al. (2003).

Methane is a potent greenhouse gas that is also produced by multiple human activities (e.g. rice and cattle farming; in both cases it's the effects of these on methane-producing microbes that are important). Pre-industrial atmospheric values are estimated to be around 720 ppb, while the 2020 value was 1,879 ppb. This increase results in some 0.52 Wm^{-2} of extra heating (the value for CO_2

over this time is 2.13 Wm^{-2}). In addition, methane gives an extra 0.3 Wm^{-2} warming from indirect effects (Lan et al., 2021). It is also possible that earlier agricultural changes (such as the rise of extensive rice growing) caused increases in atmospheric methane levels thousands of years before the industrial revolution (Ruddiman, 2005).

It seems clear that the rise in the Earth's oxygen levels appears to have been a slow process. Currently it is thought that the sustained oxygenation of the atmosphere above trace levels (the so-called great oxidation event) happened around 2.2 billion years ago. However, substantial parts of the deep ocean remained anoxic until about 0.5 billion years ago (Mills et al., 2022).

While the history of oxygen levels in the Archean and Proterozoic is problematic, the levels in the geologically recent past (the last 350 million years) have also been contentious. It seems clear that over this period oxygen levels have shown a certain amount of stability (Berner 1998; Lenton, 2001; Catling and Claire, 2005); however, there are disagreements over its extent. The classic palaeoecological approach to try to set limits to the oxygen levels in Earth's atmosphere was developed by Watson et al. (1978); they pointed out that since the development of the first forests there was a continual history of both extensive terrestrial vegetation and charcoal in the fossil record. The conclusion they drew from this was that oxygen levels had never been low enough to prevent forest fires (or there would be gaps in the charcoal record) and never so high that wet vegetation could burn uncontrollably—otherwise forests could not have existed (Figure 8.5). They attempted to quantify these oxygen levels by burning papers strips (proxies for vegetation) in different oxygen levels, concluding that oxygen abundance in the atmosphere must have stayed between 15% (dry paper would not ignite) and 25% (wet paper would burn). While the lower limit has been generally accepted, the upper limit of 25% has been controversial (e.g. Robinson, 1991; Berner, 1998).

In the early twenty-first century, new combustion experiments by Wildman et al. (2004) using more

Figure 8.5 Fire and vegetation: this photograph shows clouds of smoke billowing across Yosemite Valley in California (USA) from a large forest fire. Over Earth's history the higher the atmospheric oxygen content, the greater the chance of large fires destroying significant amounts of vegetation. On modern Earth, fire consumes so much plant material that it can be usefully compared with herbivorous animals. Indeed Bond and Keeley (2005) describe fires as effectively a 'global herbivore' in the title of a highly cited paper, which argued that the significance of fire in global ecology justifies much more ecological attention than has previously been the case. Within Yosemite National Park, fire is now viewed as a natural part of the ecology and smaller fires are often left to run their course. However, in the last few years, the National Park has been threatened by several very large wildfires—likely linked to climate change.

realistic plant material (wooden dowels and needles from white pine *Pinus stobus*) suggested that oxygen levels of 35% would not lead to uncontrollable wildfires destroying all vegetation. These experiments are obviously an improvement on using paper strips, but they also have problems in their interpretation. Using small fragments of plant material (needed to fit in their apparatus) is obviously not the same as using intact forest vegetation. Obviously burning whole forests is a difficult and ethically dubious experiment to do, and effectively impossible at oxygen levels different from the current 21%. However, it is relevant to these arguments that not all forests ignite at current levels of atmospheric oxygen, even if the vegetation is dry. During the Second World War, the German Luftwaffe accidentally carried out a fascinating series of experiments by repeatedly dropping incendiary bombs on all types of British vegetation under most conceivable weather conditions, and the deciduous woodland of England did not burn even when very dry (Edlin, 1966). Since the first edition of this book a consensus seems to have developed that fire-related arguments likely allow oxygen values of at least 30% on occasion—although in very dry climates this value may be too high (Lenton and Watson, 2011). However modelling studies by Vitali et al. (2022) suggest that while fire will suppress the extent of forest cover at very high oxygen levels it's possible that, contrary to what has been assumed in the past, there is no oxygen concentration beyond that which it is impossible for some (wet) forest to survive. However, one possible complication is that these discussions tend to assume the nitrogen levels of the atmosphere have not significantly changed—this may well be correct; however, since the mixing ratio of oxygen and nitrogen is relevant in determining flammability, any change in nitrogen would be important (Lovelock, 2006).

High oxygen levels may be associated with the ecology of the coal-forming forests. During the Permian and Carboniferous (approximately 250–360 million years ago) large coal deposits were formed, possibly due to a shortage of lignin-degrading species. As described above, the amount of oxygen entering the atmosphere by photosynthesis is largely controlled by the burial of organic material. Thus, an increased level of oxygen would be expected at this time (Robinson, 1990). Indeed this is the geological period when several lines of evidence suggest oxygen levels above 25% (Berner, 2004). C_4 photosynthesis is very rare or absent before about 7 million years ago (Retallack, 2001), so the plants of the Carboniferous are all likely to be using the standard C_3 pathway, which suffers from photorespiration. The relative rates of photosynthesis and photorespiration lead to the fractionation of carbon isotopes, so that (Berner, 2004):

$$\Delta^{13}C = a + (b - a)\, C_i/C_a$$

where

- $a = \Delta^{13}C$ which occurs through diffusion through the stomata.
- $b = \Delta^{13}C$ for the fixation by rubisco.
- C_i = intercellular CO_2 concentration.
- C_a = CO_2 concentration of the air external to the leaf.

If O_2 increases then more CO_2 will be outcompeted for access to active sites on rubisco, resulting in a higher intercellular concentration in the leaf (C_i). From the equation above it can be seen that this should lead to an increase in $\Delta^{13}C$. Experimental work by Beerling et al. (2002) showed that raising O_2 to 35% increased $\Delta^{13}C$ relative to plants grown at 21% O_2. The same authors also looked at $\Delta^{13}C$ in fossil plant material and found evidence of high O_2 levels during the Permo-Carboniferous, as predicted by the higher burial of terrestrial organic matter at this time.

Irrespective of the upper limit of oxygen (25 or 35+%), it is clear that the oxygen concentration of the atmosphere has stayed between certain limits over the last 350+ million years. Indeed Catling and Claire (2005) suggest that it has remained stable at 0.2+/−0.1 bar over the Phanerozoic. This appears surprising, since the total oxygen content of the atmosphere is cycled through the biota once every ~4,500 years—this is the time it would take for oxygen levels to drop to near zero if all photosynthesis were to stop (Lenton, 2001, 2016). If this turnover happened on a timescale of 100 million years, then this stability would not be very surprising. However, with a turnover on the shorter timescale of thousands of years, there is a need for some regulating mechanism to explain the relative stability. Since the proposed regulating mechanisms (e.g. Lenton and Watson, 2000b; Lenton, 2001) involve carbon sequestration they will be discussed in the next chapter (section 9.5).

8.5 Is photosynthesis a fundamental process?

Photosynthesis is obviously not necessary to life. Ecosystems that are not light-based are known on Earth from sites such as around hydrothermal vents and some apparently sealed caves. However, many of the organisms involved in these systems use oxygen, which is the product of oxygenic photosynthesis on other parts of the planet (Santini and Galleni, 2001). Oxygenic photosynthesis appears necessary for multicellular organisms, long food chains, and the high biomass seen on Earth. As such it is potentially very important in allowing life to have large planet-wide effects. However, it is not necessary for this, as the proposed Archean methane-dominated ecosystem would have greatly modified Earth's greenhouse effect (Box 8.2). It is possible that oxygenic photosynthesis is of such importance as a long-term source of energy to a planet's ecosystem that it could evolve on most planets with persistent ecologies. Indeed Fenchel et al. (1998, p 272) have speculated that it is so likely that 'oxygen-rich atmosphere may be an expected outcome for any terrestrial planet with an abundance of water, illuminated between about 300 and 800 nm, and with characteristics otherwise suited for organic life'. However, molecular studies have led to the general conclusion that, on Earth, oxygenic photosynthesis evolved only once (see work reviewed by Kühlbrandt, 2001 and Nisbet and Fowler, 2003).

There are two ways of viewing these molecular results. One is that oxygenic photosynthesis is very unlikely and that on Earth we were lucky; it evolved, but only once because the necessary merging of photosystems I and II is a very unlikely event. It seems likely that this came about by the coupling of two microbes: one with reaction centres related to photosystem I and the other with a version of photosystem II—see Canfield (2014) for a reasonably non-technical account. Alternatively one could take the view that such a merger is not especially unlikely but that once it has evolved on a planet there is no vacant niche for oxygenic photosynthesis, so it cannot evolve again because of the presence of competitors. An informal engineering analogy would be to point out that once the motor car had been invented it made much more sense to modify existing designs than invent the whole idea again from scratch—which would be too expensive to do with competitors already producing working vehicles. These two arguments have all the usual difficulties associated with discussions of the origin of life: while we only have one example it's difficult to say much about probabilities with any confidence!

Why is oxygen so important for a lively biosphere? It is an oxidant; indeed it is exceeded in its reduction potential by only fluorine, chlorine, and a few other chemical 'species'. Importantly, it is relatively stable under the conditions relevant to biology, which allows living systems to have control over its behaviour in chemical reactions (Fenchel et al., 1998). In addition oxygen is common in the universe; for example it is the third commonest element by mass in our solar system (after the much commoner hydrogen and helium), comprising 0.91% of the total (Gribbin, 2000).

There is a huge difficulty with answering the question 'Is oxygenic photosynthesis so likely to arise on a planet that it should be considered a fundamental ecological process?' As pointed out above, the difficulty is the same one we face in answering questions about the probability of life evolving on a planet; we only know one example—Earth—and have no idea if our planet is typical or extremely unusual! However, with oxygenic photosynthesis there is some hope that future editions of this book could be revised in the light of real data. We don't need to visit a planet to find oxygenic photosynthesis (while a visit would be necessary to find *small* amounts of microbial life) as we can potentially detect oxygen-rich atmospheres remotely from Earth (Lovelock, 1965; Sagan et al., 1993; Kasting, 2010). However, the identification of oxygenic photosynthesis on another planet would be such an important discovery that it may prove difficult to establish to everyone's satisfaction that the oxygen *must* have come from photosynthesis, rather than some other unknown abiological mechanism.

While currently it is difficult to say anything definite about the likelihood of photosynthesis (or indeed life!) on other planets, it is possible to theoretically investigate the conditions that could lead to photosynthesis and the production of oxygen in quantities large enough to be detected by Earth-based astronomers. For example, Wolstencroft and Raven (2002) attempted to model the factors that may limit—or allow—oxygenic photosynthesis on Earth-like planets. Their model suggests different levels of photosynthetic oxygen generation, depending on the type of star the planet was orbiting. This could aid the selection of Earth-like planets for detailed spectroscopic study once the technology is available.

8.6 Overview

Photosynthesis—both anoxygenic and oxygenic—allows access to new sources of energy on a planet. Oxygenic photosynthesis has the potential to create an oxygen-rich atmosphere and so allow aerobic respiration, which yields much higher amounts of energy than anaerobic respiration (fermentation) and so allows longer food chains and more complex ecosystems. The amount of oxygen added to the atmosphere is intimately linked to the burial of organic matter in sediments, and (contrary to the impression given in many ecology books) marine phytoplankton are crucially important in maintaining the levels of atmospheric oxygen on Earth.

Anoxygenic photosynthesis will have a positive Gaian effect by providing a new and important source of energy, thereby increasing the probability of the long-term survival of life on the planet. Oxygenic photosynthesis is more problematical; it could potentially have either a positive or a negative effect. As with anoxygenic photosynthesis, it provides an important energy source, but the oxygen given off is likely to be toxic to organisms evolved in anoxic conditions (Fenchel and Finlay, 1995). Could such 'oxygen pollution' kill off all life on a planet (perhaps by destroying some crucial guild of anaerobic microbes or, in the case of early Earth, destroying the methane greenhouse)? If so, oxygenic photosynthesis has the possibility of behaving in a negatively Gaian manner. However, it is possible that these ecological problems associated with the creation of an oxygen-rich atmosphere may be less serious than is sometimes assumed. Oxygen levels in the atmosphere rose very slowly, over timescales many orders of magnitude slower than generational times of simple organisms. Clearly the photosynthetic organisms themselves must have evolved tolerance to oxygen; the same is presumably true for organisms that lived in close proximity to the oxygen producers (e.g. feeding on the surface of photosynthetic microbial mats). The genetic variation required to deal with oxygen-rich conditions is likely to have been widely available via horizontal gene transfer long before global oxygen levels started to rise. In addition, we should remember that Earth still has lots of species of anaerobic microorganisms; they are just restricted to oxygen-free habitats such as some waterlogged sediments or the guts of many animals (Fenchel and Finlay, 1995).

At the current state of our knowledge it is impossible to know if we should expect most biospheres to evolve oxygenic photosynthesis. However, improvements in telescope technology should allow us to look for oxygen-rich atmospheres around distant Earth-like planets and may help answer this question over the next few decades.

CHAPTER 9

Carbon sequestration

9.1 Carbon sequestration and landscape change in northwest England

During the eighteenth century in much of Europe there were significant changes in the approach to knowledge (as well as moral and social values) which slowly started to appear more modern in their outlook. Between 1724 and 1727 Daniel Defoe—best known today as the author of *Robinson Crusoe*—published his *Tour Thro' the Whole Island of Great Britain*, describing, in three volumes, a journey around the country in which he praised the enterprise, commerce, and industry of eighteenth-century Britain (Porter, 2000). Chat Moss, a 2,450-ha expanse of peat bog between Liverpool and Manchester, did not meet with his approval. He wrote: 'The surface at a distance, looks black and dirty, and indeed frightful to think of, for it will bear neither horse or man, unless in an exceedingly dry season, and then not so as to be passable ... what nature meant by such a useless production 'tis hard to imagine; but the land is entirely waste except ... for the poor cottagers fuel [i.e. the use of dried peat as a fuel source], and the quantity used for that is very small'.

Chat Moss today is very different (Figure 9.1). Drainage of these boglands started a few decades after Defoe was writing, and in the early 1760s the Duke of Bridgewater extended his recently constructed canal onto the northern part of the bog, both to help drain the land and to allow spoil from his coal mines to be dumped there. By the end of the eighteenth century parts of the most eastern area of the bog had been leased with the aim of draining it and turning it into agricultural land. This campaign of reclamation escalated when the Liverpool-to-Manchester railway opened in 1830, crossing the bog—a considerable engineering feat, because of the very wet and unstable ground conditions—making access for reclamation much easier (Wilkinson and Davis, 2005). Today most of the area is drained and used for agriculture and until recently commercial peat extraction, with a small area of less-modified bog that has been managed for nature conservation.

In 2013 the Lancashire Wildlife Trust started to restore areas that had been used for peat extraction, with the intention of returning the site to an active peat-forming bog (Osborne et al., 2021). As peat is largely composed of the incompletely decomposed remains of plants (Figure 7.4), and so is effectively temporarily sequestered carbon, this loss of peat from agriculture and commercial extraction will have returned CO_2 fixed by photosynthesis over thousands of years from the peat to the modern atmosphere. Because of this there is now a large interest in conserving remaining peatlands and restoring damaged ones to functioning bog ecosystems, as at Chat Moss. The concerns about climate change have completely changed attitudes to peatland systems during my lifetime. For example Moore and Bellamy (1974), in what was for several decades a very influential textbook on peatland ecology, made no mention of peatlands in the context of future climate changes but did have a whole chapter on the potential of the exploitation of peatlands for energy and agriculture. Such a book would be very different if written today—as shown by Rydin and Jeglum's (2013) *The Biology of Peatlands*, whose final chapter is on 'Peatlands and climate change'.

This example of peatlands illustrates the potential for life to sequester carbon in sediments—something that I will argue is fundamental to any

The Fundamental Processes in Ecology. Second Edition. David M. Wilkinson, Oxford University Press.
 DOI: 10.1093/oso/9780192884640.003.0009

Figure 9.1 Part of Chat Moss, once a large area of peat bog between Manchester and Liverpool in northwest England; it has now been reclaimed for agriculture and other uses. Parts are undergoing restoration back to functioning peatland habitat. The historic reclamation for agriculture led to breakdown of the peat and release of carbon dioxide back into the atmosphere. The photograph (taken in 2004) shows an area of extensive commercial peat cutting, for horticultural use. This forms part of the area that is currently undergoing restoration with the aim of re-establishing functioning peat bog.

Earth-type ecological system and is a process of great practical importance in the context of 'global warming' on Earth. One of the objectives of this chapter is to give a longer timescale Earth systems context to this crucial problem—a context that is usually missing from the standard ecology texts.

9.2 A tale of two cycles: short- and long-term carbon cycles

Look in most university-level ecology textbooks and there is likely to be a diagram labelled 'the carbon cycle', and as most of these books are organized around a hierarchical series of entities, this planet-scale system will probably be towards the back of the book. Although often described as *the* carbon cycle, such diagrams are usually depicting only a small part of the planet's carbon cycle, often referred to as the 'short-term carbon cycle' by those with a more geological perspective on these things (e.g. Berner, 2004; Kump et al., 2010). These ecology texts usually either miss out the two-way fluxes of carbon between rocks and the surficial system (i.e. oceans, atmosphere, life, and soils) or restrict it to considering the burning of fossil fuels. This is a reasonable approximation on the timescale of thousands of years; however, this is a tiny fraction of the time span of life on Earth. To put the role of life in its proper Earth systems context we need to consider the 'long-term carbon cycle'. A generalized mass balance equation for the long-term cycle can be written as (Berner, 2004):

$$dM_c/dt = F_{wc} + F_{wg} + F_{mc} + F_{mg} - F_{bc} - F_{bg}$$

See Table 9.1 for a definition of these terms and Figure 9.2 for a graphical representation of these fluxes. The main fluxes in this equation that are described in most ecology texts are the movement of organic carbon into (F_{bg}) and out of (F_{wg}) organic sediments, along with the release of carbon dioxide from volcanoes (F_{mc}). In the case of the peatland described in the introduction to this chapter (section 9.1), the key factors were peat formation (a short-term version of F_{bg}) and the breakdown of peat sediments after drainage (F_{wg}).

Two key factors in the long-term, or geological, carbon cycle are the fluxes of carbonate carbon (F_{wc} and F_{bc}); these are often intimately connected with the weathering of silicate minerals and are usually overlooked by descriptions of the carbon cycle in ecology textbooks. The main processes in the long-term carbon cycle can be summarized in two chemical equations (Berner, 1998; Watson, 1999). The first of these is the familiar equation for oxygenic photosynthesis (or respiration if read from right to left) described in chapter 8:

$$CO_2 + H_2O \leftrightarrow (CH_2O) + O_2$$

Table 9.1 Important variables in the long-term carbon cycle (following Berner, 2004).

M_c = Mass of carbon in surficial system.
F_{wc} = Carbon flux from weathering of Ca and Mg carbonates.
F_{wg} = Carbon flux from weathering sedimentary organic matter.
F_{mc} = Degassing from volcanism, metamorphophism, and diagenesis of carbonates.
F_{mg} = Degassing from volcanism, metamorphophism, and diagenesis of organic matter.
F_{bc} = Burial flux of carbonate carbon in sediments.
F_{bg} = Burial flux of organic carbon in sediments.
F_{wsi} = Carbon flux from weathering Ca and Mg silicates due to transfer of atmospheric carbon to Ca and Mg carbonates.

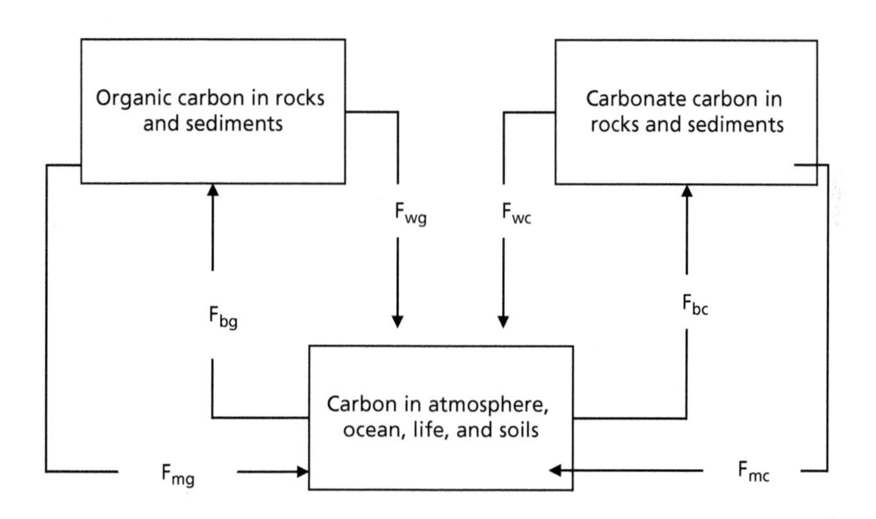

Figure 9.2 Schematic representation of the long-term carbon cycle (after Berner, 1998, 2004). See Table 9.1 for description of the various fluxes. Most of the 'short-term carbon cycle' usually illustrated in ecology textbooks is concerned with movement of carbon within the box labelled 'Carbon in atmosphere, ocean, life, and soils'. Note that the sizes of the three boxes are misleading—there is much more carbon locked up in rocks than in the atmosphere, etc. If the atmosphere, etc. box was drawn to scale it would be so small no text would fit in it. This size difference makes the equation $dM_c/dt = 0$ really remarkable. Even a very small percentage difference in flux (movement) between a very large pool and a small one should lead to a major change in size of the small pool. If this point isn't immediately obvious think of an analogous situation where a banking error causes a small percentage of the wealth of a billionaire to be moved into your bank account.

In this context it should be read as an equation for 'net photosynthesis'—that is photosynthesis minus respiration—which is the amount of organic matter buried in sediments.

The second key equation is that for the weathering of silicate minerals, namely:

$$XSiO_3 + CO_2 \leftrightarrow XCO_3 + SiO_2$$

where X stands for either Ca or Mg. Read from left to right this equation symbolizes the uptake of CO_2 from the atmosphere during the weathering of silicate minerals and its transfer to HCO_3^-, which is washed into the oceans where it is deposited as $CaCO_3$ or $MgCO_3$. Read from right to left the equation summarizes the breakdown of these carbonate minerals at depth within Earth and the subsequent release of CO_2 back to the atmosphere (Berner, 1998, 2004).

The carbon storage capacity of the sediments is far greater than that of the surficial system. As such, on a timescale of millions of years the flux of carbon between these two systems must be in a steady state (Berner, 2004), so that:

$$dM_c/dt = 0$$

Even a minor departure from this would, over geological time, lead to colossal changes in the CO_2 content of the atmosphere. In terms of the silicate weathering system described above, it must be the case that (Berner, 2004):

$$F_{wsi} = F_{bc} - F_{wc}$$

In the context of this book it is obviously of interest to ask what, if anything, is the role of ecology in the long-term stability of this system?

9.3 The role of life: from geochemical cycles to biogeochemical cycles

Feedback loops leading to regulation do not necessarily require life as part of the system, as is demonstrated by the example of the natural nuclear reactors at Oklo (section 7.5). In the case of silicate weathering, Walker et al. (1981) pointed out a potential abiotic negative feedback mechanism that could stabilize Earth's surface temperature over geological time. They suggested that if the partial pressure of carbon dioxide (ρCO_2) in the atmosphere increased, this would raise the surface temperature of the planet. This in turn would both directly increase the rate of silicate weathering (as most chemical reactions increase in rate with temperature) and indirectly increase weathering through greater evaporation from the oceans, leading to increased rainfall. Increased silicate weathering would, over geological time, reduce the atmospheric levels of CO_2. A decrease in CO_2 reduces the surface temperature through a reduced greenhouse effect, so there is a mechanism for the regulation of surface temperature on Earth that would apply even if life had never evolved (or arrived) on the planet. It's important to note that regulation is a term with many assumptions behind it, which has led to some confusion in the Earth systems science literature (Box 9.1). However, to an ecologist it usually implies 'control according to defined rules or principles' (Calow, 1999). This definition certainly includes Walker et al.'s (1981) mechanism as an example of regulation, even though Earth's temperature has not been static over geological time (Kump et al., 2010).

Box 9.1 What is regulation?

The regulation of population size through density-dependent processes (section 3.4) has been a major topic in population ecology. However, it has historically proved remarkably difficult to unambiguously demonstrate this process in wild populations (Sinclair, 1989). One of the most straightforward approaches is to experimentally alter the size of a population and see if it tends to return to its pre-perturbation size. However, this is often impracticable and we are left with the statistical analysis of time series data to search for regulation.

There are similar problems in the ecology of the Earth system (e.g. Volk, 2004). A crucial question is, 'What would we expect to see if oxygen or carbon dioxide was regulated over geological time?' Some people clearly expect that a regulated system should stay the same from time period to time period, so (for example) oxygen content of the atmosphere should stay constant over geological time. However, we know from population ecology that regulated processes can show much more complicated behaviour. Consider the difference equation analogue of the differential logistic equation, key to much of theoretical population ecology:

$$N_{t+1} = N_t \exp[r(1 - N_t/K)]$$

where N is population size (at time t or t+1), r is the population growth rate, and K is the carrying capacity. Classic studies by Robert May (1974) showed that by changing the value of r, a diversity of outcomes were possible. These ranged from a stable population to a cyclical one or even chaotic behaviour, although this equation clearly has regulation around 'K' built into it.

A more intuitive feel for these problems can be gained by thinking about heating systems regulated by a thermostat, such as the heating/air conditioning system in a building or a hot-water shower. Ideally you would want the temperature of the air or water to stay at exactly whatever temperature the thermostat had been set to. However, stand under an old hot-water shower and you may feel the water getting hotter, then colder, then hotter again as the water temperature fluctuates around the set point; the thermostat is still regulating the temperature although it's not staying constant with time (the record for Earth's surface temperature through the glaciations of the last 2 million years shows an intriguingly similar pattern).

An additional problem with the idea of regulation is that merely saying that a system is 'regulated' misses out important information about how the 'thermostat' is set. Consider a house: if you are told that the heating is well regulated then this gives you important information about conditions in the house—you would normally expect that the temperature will not change much over time irrespective of what happens to the temperature outside. However, some important information is missing; you would not know if you needed lots of warm clothes in the house or if you could comfortably walk around naked unless you have been told what temperature the thermostat is set at. One of the great strengths—and provocations—of Gaia theory is that it not only suggests large-scale regulation involving life, but also specifies that this will be for a 'habitable state for whatever is the current biota' (Lovelock, 2003, p 769).

This box may appear to be a long digression from the topic of this chapter, but my experience in discussing 'regulation' in an Earth systems context over the last few years suggests that it is a concept that is prone to much confusion.

Walker et al.'s (1981) model was purely abiotic; however, Lovelock and Watson (1982) quickly pointed out that terrestrial life was probably crucial to a full understanding of the effects of silicate weathering on atmospheric CO_2 and climate. In the absence of terrestrial life, it is likely that soil would be rare due to erosion by water and wind action. With the first terrestrial 'vegetation', such as a stabilizing crust of bacteria, algae, and possibly lichens (see Figure 1.2), soils would be able to start to resist erosion (Schwartzman and Volk, 1989). This simple story is complicated by the existence of a number of Archean palaeosols, dating back to around 3,500 million years ago (Retallack, 2000), the general assumption being that this is long before any life on land—but see Schwartzman (1999) and Lenton and Daines (2017) for suggestions of limited life on land 3 billion or so years ago. Clearly the—much later—evolution of vascular plants will have allowed much deeper soils to develop.

Biology increases the rate of silicate weathering by a number of mechanisms (Lovelock and Watson, 1982; Schwartzman and Volk, 1989; Berner, 2004; Zambell et al., 2012). For example, the respiration of soil-living microbes increases ρCO_2 in the soil and the plants and their associated mycorrhizae produce a range of organic acids. In addition, a soil stabilized by life provides a much greater surface area over which chemical weathering can take place; Schwartzman (1999, p 51) calculated that a hypothetical soil 1 m deep composed of particles that were cubes of 1 mm^2 (for ease of calculation) has 6,000 times the potential reactive surface of an impermeable bare rock surface. At a global scale, Porada et al. (2014) estimated the combined effects of lichens and bryophytes on weathering rates and suggested a figure for chemical weathering of between 0.058 and 1.1 km^3/year (Figure 9.3).

These arguments suggest that the first terrestrial 'vegetation' could have had important effects on the rate and/or extent of silicate weathering, and hence carbon sequestration. It therefore seems interesting to ask, 'When was the first life on land?' Just as with the origin of life on Earth or the age of the first eukaryotes, this turns out to be a difficult question to answer. Traditionally palaeontologists have thought that land plants evolved during the Devonian, with little terrestrial vegetation before this time. However, it now seems likely, mainly based on evidence from fossil spores, that there was some terrestrial vegetation during the Ordovician (Kenrich and Crane, 1997; Retallack, 2000; Beerling, 2019a). The extent of this vegetation is currently unclear, with many authors still suggesting a late Silurian/early Devonian timing for *extensive* terrestrial vegetation (e.g. Kenrich and Crane, 1997; Beerling et al., 2001; Beerling, 2019a). Edwards et al. (2015) suggest that this Ordovician vegetation may not have been extensive enough, or had deep enough roots, to have had a major effect on carbon burial. Prior to colonization of the land by true plants, the general assumption is that there would have been terrestrial microbial crusts since the late Proterozoic (Kenrich and Crane, 1997). However, such life may be much older: for example over 20 years ago, Retallack (2001, p 200) concluded that 'the idea of life on land as far back as 3500 [million years] is no longer outrageous speculation ... indeed there is isotopic and geochemical evidence for a variety of microbial ecosystems in Precambrian paleosols'. Because any such life is likely to have been in the form of thin microbial mats, which do not easily fossilize, indirect evidence from the potential biogeochemical effects of early life on land becomes an important alternative line of research. There is indication from weathering that such mats may have existed back to even c.

Figure 9.3 Bryophytes (in this case mosses) and lichens on a granite rock in Wistman's Wood, Dartmoor, southwest England. Bryophytes and lichens have multiple impacts on global biogeochemical cycles, both today and in the past (Porada et al., 2014). For example, they affect weathering rates (see main text) and also nitrogen fixation, through associations with cyanobacteria. Some lichen species contain nitrogen-fixing bacteria as symbionts, while most mosses are colonized by nitrogen-fixing bacteria. Although the presence of such bacteria in mosses has been recorded for over 100 years (e.g. Warming, 1909) it has only recently become apparent that this is an ecologically important process. Nitrogen fixation associated with bog moss *Sphagnum* spp. can be several orders of magnitude larger than those associated with most other moss types—such as those seen in this photograph (Rousk, 2022).

3.0 billion years ago. Key evidence includes oxidative weathering (interpreted as localized high levels of oxygen from photosynthetic microbes), and after the great oxidation event the mobilization of phosphorus, iron, and other elements in weathering profiles (Lenton and Daines, 2017).

While the history of the early terrestrial 'vegetation' is uncertain, by the mid Devonian there is clear evidence for forest soils (Retallack, 1997, 2001). The arguments of Lovelock and Watson (1982) suggested that such well-developed soils should have a potentially important effect on atmospheric CO_2 via their effects on silicate weathering. At the end of the 1980s Tyler Volk and David Schwartzman (Volk, 1987; Schwartzman and Volk, 1989) formally modelled these effects, showing that they were indeed important, the size of the effect on atmospheric CO_2 depending on the exact values used to quantify the biotic enhancement of weathering. Schwartzman and Volk (1989) calculated that if this biotic enhancement of weathering was 10, 100, or 1,000 times the abiotic rate, the current temperature of an abiotic Earth would be 15, 30, or 45°C warmer than currently observed. As they pointed out, the two higher figures would tend to give a current surface temperature above that which can be tolerated by most eukaryotes (Table 1.1). It seems that on Earth the biotic enhancement of weathering may have been crucial to the continued survival of eukaryotes. As the sun's energy has increased (Sagan and Chyba, 1997), the CO_2 greenhouse has decreased, helping to regulate Earth's surface temperature. However, work by David Beerling, Robert Berner, and colleagues (described in the next section) has complicated this interpretation by suggesting a range of new *positive* feedback loops involving plants and CO_2.

Life also plays a profound role in the global carbonate cycle through the influence of marine plankton. On a geological timescale marine sediment can be a major sink for carbon, in carbonate form (either as calcium or as magnesium carbonate). The role of life in this system became particularly important during the Mesozoic with the evolution of calcareous microplankton along with a general diversification of planktonic communities (Rigby and Milsom,

2000). The precipitation of carbonate is described by the following summary equation (Ridgwell and Zeebe, 2005):

$$Ca^{2+} + 2HCO_3^- \rightarrow CaCO_3 + CO_2 + H_2O$$

This process is surprisingly counter intuitive in its effects on carbon sequestration. On the timescale of interest to most geologists, it causes a removal of carbon from the oceans and atmosphere by the addition of carbon to marine sediments (F_{bc} in Figure 9.2). However, on the much shorter timescales which usually dominate ecological research, there is an increase in CO_2 in the ocean and atmosphere (Ridgwell and Zeebe, 2005).

9.4 Co-evolution of plants and carbon dioxide on Earth

Carbon dioxide is a key raw material for photosynthesis. As such, it seems likely that changes in its atmospheric concentration will affect plants. As I have described above, the presence of land plants also had a big effect on CO_2 concentration, so we have a fascinating feedback system between plants and CO_2. When plants moved onto land, water loss would have become a problem—which is obviously not the case for aquatic plants. In response to this, they evolved a cuticle that is relatively impervious to water. However, photosynthesis requires that CO_2 can enter the photosynthetic tissues of the plant, so this cuticle has to be pierced by stomata (pores that allow air to enter the plant but also water loss from the plant: Givnish, 1987; Grace, 1997). There is a trade-off for plants between conserving water and accessing CO_2. From this it follows that the lower the CO_2 content of the air, the greater a terrestrial plant's potential problem with water loss, as it will need more stomata to access CO_2. For example, preserved leaves of the dwarf willow *Salix herbacea* from sediment samples from the last glacial/interglacial cycle show stomatal densities increasing during low CO_2 glacial periods (Beerling et al., 1993). Globally, huge amounts of water are now transpired from plants each year through their stomata, with something like 40% of all rainfall on land coming from water that has entered the atmosphere via transpiration by plants. In David Beerling's (2019a, p 100) evocative words, 'plants are really upside-down waterfalls, showering the atmosphere with water vapour that eventually returns to Earth as rain'. In so doing they make for a wetter planet, more suitable for extensive terrestrial vegetation (Kleidon et al., 2000; Berry et al., 2010).

One of the surprising aspects of the palaeobotanical record is that there is a gap of at least 40 or 50 million years between the origin of vascular land plants and the first large leaves (Kenrich and Crane, 1997; Beerling, 2005). Plants photosynthesize, and leaves are effectively their solar panels, so it seems obvious that leaves would quickly evolve to allow plants to capture more light. A plausible explanation for this delay links CO_2 levels, stomatal density, and the heat budgets of plants (Beerling et al., 2001; Beerling, 2019a). CO_2 fell during the later Palaeozoic as terrestrial vegetation accelerated the role of silicate weathering and enhanced carbon burial (described in section 9.3). This reduction in CO_2 caused an increase in stomata and so increased transpiration, having the effect of allowing plants to keep large leaves cool—although the initial selection pressure was for access to declining CO_2. Today plants can (on occasion) maintain leaf temperatures below air temperature due to the cooling effect of transpiration (Monteith, 1981). To illustrate this, consider these data collected on a sunny day in May in south-west England when the air temperature was 15°C; while grey paper in direct sunlight was 30°C and black slate roof slates were 54°C, sunlit leaves of several species of trees were only 14°C—such is the cooling power of transpiration (James Lovelock, pers. comm.). If the leaves had been at temperatures around 40°C then photosynthesis would have stopped, and at the temperature of the roof slates they would have died. The cooling effect of transpiration is even more strongly illustrated by the dark-red leaves of some trees in autumn. There has been a long-standing idea that such leaves may be able to maintain higher temperatures than green leaves, in the cool of autumn, because these darker pigments absorb more solar energy (Wheldale, 1916; Wilkinson et al., 2002). However, detailed measurements in New England, USA suggest that these dark colours have no effect on temperature (Lee et al., 2003), presumably because of the effect of transpiration.

Modelling studies suggest that a plant with large leaves but the low stomatal densities of early land plants would have been at risk from death from overheating (Beerling et al., 2001a). As such it can be argued that plant responses to low CO_2 (evolving higher stomatal densities) allowed large leaves to develop (Beerling, 2005). However, this raises the question of why couldn't increased stomatal density evolve as an adaptation for regulating the plant's temperature, even at high ρCO_2?

One of the interesting implications of these ideas of David Beerling and colleagues is that many of the feedbacks between terrestrial vegetation and CO_2 over geological time may be positive, and so potentially destabilizing (Beerling and Berner, 2005). For example, the development of larger land plants is made possible by lower CO_2 levels and the presence of this vegetation further reduces CO_2 in the atmosphere through their photosynthesis. It is currently very difficult to quantify the overall effect of terrestrial vegetation in this context and to quantitatively compare the relative effects of negative feedback processes (Lovelock and Watson, 1982; Schwartzman and Volk, 1989) with the positive feedback processes (Beerling and Berner, 2005).

9.5 Oxygen and carbon sequestration on Earth

As described in the previous chapter (section 8.4), the paired summary equations of oxygenic photosynthesis and aerobic respiration illustrate the close links between the carbon and oxygen cycles in the Earth system. They are so important that I repeat them here:

$$CO_2 + 2H_2O \rightarrow (CH_2O) + O_2 + H_2O \text{ [photosynthesis]}$$

$$(CH_2O + O_2 \rightarrow CO_2 + H_2O) \text{ [respiration]}$$

Biochemically, aerobic respiration is the reverse of photosynthesis, so for oxygen produced by photosynthesis to persist in the atmosphere some organic material must escape respiration by other organisms and become buried in sediments. Without such carbon sequestration, oxygen cannot build up in the atmosphere.

The continuous charcoal record suggests limited variation in oxygen levels over the last 350 million years (Lenton and Watson, 2000b; and see section 8.4 of this book). The link between the carbon and oxygen cycles means that aspects of the carbon cycle have been suggested to play important roles in the regulation of atmospheric oxygen. For example, the rate of silicate weathering under the influence of vegetation and CO_2 has important implications for ρO_2. This is because increased weathering causes more phosphate—from rocks—to enter the oceans, so allowing more planktonic photosynthesis. This photosynthesis produces a net increase in atmospheric oxygen, if some of the planktonic carbon is sequestered in marine sediments. At the same time high ρO_2 increases the frequency and magnitude of forest fires, providing a negative feedback to the silicate weathering cycle by reducing terrestrial vegetation (Lenton and Watson, 2000b; Lenton, 2001; Bergman, et al., 2004). The long-term sulphur cycle is also linked to the carbon and oxygen cycles: for example by the oxidative weathering of pyrite FeS_2 (Berner, 2004):

$$15O_2 + 4FeS_2 + 8H_2O \rightarrow 2Fe_2O_3 + 8SO_4^{-2} + 16H^+$$

The tendency for ecology textbooks to often consider these cycles as isolated systems, which can be described independently, is obviously a gross simplification at these more geological timescales. However, the central role of life in these long-term cycles should make them of great interest to ecologists.

9.6 Humans and carbon sequestration

The preceding discussion highlights the role of life in carbon cycling, and currently one species, *Homo sapiens*, is having unprecedented effects on this system (Figure 9.4). In a widely read popular survey of the state of the life sciences at the end of the 1920s Wells et al. (1931; much of the text was written by the biologist Julian Huxley) in describing the carbon cycle wrote of fossil fuels that: 'The fireplace, the factory, and the auto mobile are doing all they can to restore this deposited carbon to a state of gaseous accessibility'. Unlike such a book written today, they made no mention of the potential effects of this on the global climate (but did discuss the adverse effects of pollution from burning coal on human health). Things have changed and now

Figure 9.4 Winding wheel at Astley Moss Colliery, northwest England—for lowering miners down the mine shaft and bringing coal back to the surface. This was one of many coal mines that developed in Britain during the nineteenth and twentieth centuries, and is now a museum. The problem of how to pump water out of British coal mines led, during the eighteenth century, to development of steam engines to power the pumps. At first these were extremely inefficient; however, as they were situated at the pit head coal was easily and cheaply available, so these inefficiencies were not so great as to make them uneconomic. With this growing use of its fortuitously accessible coal and steam engines, 'England underwent an industrial revolution that changed everything, not just for England but also for the world' (Marks, 2020, p 119).

the climatic effects of burning fossil fuels, and other changes we are making to the Earth system, are a widespread, urgent, and high-priority concern. In Tyler Volk's (2008, p ix) view, the 'consequences will sweep all portions of the globe ... The rise in CO_2 will either help unite the emerging global society to counteract a shared alarm or will divide it into competing camps'. Since he wrote these words a 'global society' looks increasingly fragmented and 'competing camps' significantly more likely.

This is not a book on applied ecology; however, it is impossible to write a chapter on carbon sequestration without any mention of the effects of humans on the future of the Earth system. With 'global warming' (or 'global heating' a better term to describe the physics of the system as the greenhouse gases heat the Earth—Wilkinson, 2021a) we are not taking part in any new processes within the Earth system, but we are greatly altering the rate of some of the existing processes (see Box 9.2 for a brief history of human intervention in the Earth's carbon cycle). In the context of Figure 9.2 we have greatly increased the rate of F_{wg} by burning fossil fuels. In addition, we have altered the distribution of carbon within the 'surficial system' box by processes such as biomass destruction—which converts carbon from organic matter into atmospheric CO_2. In the context of the ideas discussed so far in this chapter, it is interesting to ask if we could perhaps replicate some of the changes that happened with the rise of land plants to help draw down CO_2 levels in the atmosphere (Beerling, 2019b). At the very least we could stop reducing forest cover and start to increase it again, although there are multiple complications—such as the effect of albedo change and the fact that area under forest doesn't always translate into increased CO_2 storage (the type of forest and its management can be very important—Naudts et al., 2016).

Box 9.2 Chronology of events relating to humans and the carbon cycle (based, in part, on Grace, 2004 and Dyke, 2021).

Before 1800: For thousands of years human activity, such as forest clearance, has probably caused changes in the level of greenhouse gases (e.g. carbon dioxide and methane) in the atmosphere.

1780-1820: Industrial Revolution; a dramatic increase in the use of coal and so a rise in atmospheric carbon dioxide (Figure 9.4). One possible date for the start of the Anthropocene is 1712, the year Thomas Newcomen built his early steam engine (Lovelock, 2014).

1859: Oil wells sunk in Pennsylvania, USA; oil is soon being produced in many countries.

1896: Swedish chemist Svante Arrhenius suggests that increasing emissions of carbon dioxide will lead to global warming.

1903: Ford motor company founded in the USA; this quickly leads to the start of the mass production of cars.

1958: Charles Keeling starts measuring atmospheric carbon dioxide on Mauna Loa in Hawaii using an infrared analyser; these measurements are still being made today (Heimann, 2005).

1968: First photographs of Earth from deep space. These have a dramatic effect on how people view Earth; it looks a lot smaller and easy to damage when seen from space. The astronomer Fred Hoyle (1950) predicted the psychological effect of such photos on our view of the planet almost 20 years before they were taken!

1979: James Lovelock publishes his first book on *Gaia*, suggesting that life may be crucial to the regulation of Earth's climate and chemistry.

1987: Ice core data from Antarctica show a close correlation between carbon dioxide and temperature over the last 100,000 years.

1988: Intergovernmental Panel on Climate Change (IPCC) is established.

1997: Kyoto Protocol, an international agreement to limit greenhouse gas emissions.

2005: Kyoto Protocol formally came into force in February, but without the USA.

2015: Paris Agreement sees 195 countries commit to keeping warming to below 2°C.

2020: An air temperature of 54.4°C is recorded at Furnace Creek in Death Valley, USA on 16 August—the highest temperature ever recorded. The decade 2010–2020 is confirmed as the warmest decade ever recorded since records began.

While silicate weathering feedbacks control pCO_2 on geological timescales, it has little effect on timescales less than 100,000 years (Ridgwell and Zeebe, 2005)—so while it will eventually help to reduce CO_2 artificially increased by human activity it will not do so on a timescale considered useful by most people. This raises an interesting question: 'Is there any way of speeding up this feedback to make it of use on much shorter timescales?' One possibility is to add crushed silicate-rich rocks (e.g. basalt) to agricultural soils (Beerling et al., 2020). The powdered rock would have a high surface area to volume ratio and so weather much more quickly than larger rocks. Such an approach would also potentially have benefits to soil nutrient cycles. However, attempts to model the effectiveness of this idea show it can only be at best a partial solution to the problem of high atmospheric CO_2 levels, and it is possible that the soils that would be the most practical to use may not be those best suited to rapid silicate weathering. On shorter timescales than the silicate weathering cycle, the marine carbonate cycle does play a role in atmospheric CO_2 and this is currently a cause for concern, as increased atmospheric CO_2 causes more CO_2 to dissolve in the ocean. This lowers the ocean's pH and decreases CO_3^{2-}—one of the building blocks of calcium carbonate—according to the following equation (Gattuso and Buddemeier, 2000):

$$CO_2 + CO_3^{2-} + H_2O \rightarrow 2HCO_3^{-}$$

This can cause decreased calcification of plankton (Riebesell et al., 2000). It is currently uncertain how a decrease in calcified plankton will feedback to ρCO_2 in the atmosphere. One possibility is that a reduction of carbonate production in the surface ocean would reduce the amount of CO_2 in the water. This would allow the water to absorb more atmospheric CO_2 and so act as a negative feedback on global warming (as carbonate production releases CO_2 into the water, as described at the end of section 8.4). However, it is also plausible that carbonate production helps remove particulate organic carbon from surface waters to ocean sediments, so less carbonate production could reduce the removal of organic carbon to ocean sediments and so act as a positive feedback on global warming (Ridgwell and Zeebe, 2005). The size of the ocean carbon sink is an area of ongoing research; however, the more CO_2 taken in by the ocean, the larger the potential problem with ocean acidification (Terhaar et al., 2022).

Probably the first direct human intervention in the short-term carbon cycle was forest clearance. For example, 5,000 years ago the dominant vegetation types over most of Britain were various sorts of forest, ranging from pine *Pinus sylvestris* and birch *Betula* spp.-dominated forests in the north to lime *Tilia* spp.-dominated forest in the southeast of England and oak *Quercus* spp.-dominated ones over much of southwest England (Bennett, 1989; Birks, 2019). As agriculture developed most of this forest was lost, a trend that has been repeated in many parts of the world. Agriculture first appeared around 12,000 years ago at the eastern end of the Mediterranean and soon after in China, although just as with the ages of the first life or first eukaryotes there are disagreements about the details of exact timing (Lev-Yadum et al., 2000; Cunliffe, 2015). Over the last 12,000 years the idea of agriculture has been invented on a minimum of five occasions around the world—and possibly on nine or more independent occasions (Diamond, 1997, 2002). William Ruddiman (2003, 2005) suggested that deforestation associated with agriculture was affecting global CO_2 levels by around 8,000 years ago and methane production (from rice paddies, etc.) rose around 5,000 years ago. These dates have proved controversial and recently Ruddiman et al. (2020) have published a revised version. Although things are clearly somewhat more complex than Ruddiman originally suggested, there seems to be considerable evidence in favour of the basic idea—with the evidence for a climate effect from methane from early rice cultivation perhaps stronger than that for the effects of early forest clearance.

Since the removal of forests has been a source of CO_2, replacement of forests could act as a CO_2 sink. For example, Rowantree and Nowak (1991) estimated that the urban trees of the USA sequestered around 6.5 million tons of carbon per year (as well as providing other benefits such as shade). Provided rainfall levels are still adequate, forests can cover a denuded landscape remarkably quickly (Figure 9.5). The classic example of this are the forests of eastern USA and Canada, much of which was cleared during the eighteenth and nineteenth centuries to make room for European-style agriculture. However, during the mid-nineteenth century, as people moved from farms to seek their fortunes in the newly expanding cities, much of this land went out of agricultural production and quickly returned to forest (Foster, 1992). While the North American example is the most well studied, giving rise to many detailed studies of 'old field succession' described in the ecology literature, similar processes have happened in many other places. For example, between 1960 and 1980 large areas of forest were cleared in the Chorotega region of Costa Rica—this area includes the Santa Rosa National Park, well known to ecologists through decades of groundbreaking work by Daniel Janzen and colleagues. Much of the forest was cleared for beef production. However, when, in the 1980s, beef prices fell, large areas were allowed to revert to forest (Arroyo-Mora et al., 2005). Both the North American and the Costa Rican examples are of areas that have returned to forest after earlier human-caused deforestation. However, some areas have never had forest because they were too isolated for trees to arrive and have not had long enough for the tree growth-habit to evolve from scratch, a fascinating example being Ascension Island in the tropical south Atlantic. Much of the island is arid; however, on Green Mountain there is much higher rainfall and considerable occult precipitation with water settling out directly from mist. Through human introductions of a large number of plant species, the mountain has gone from a treeless system in

Figure 9.5 Examples of new or regrown forests. Figure 9.5a Regenerating forest in southern Ontario, Canada. This is typical of much of eastern USA and Canada in having been largely cleared of forest for agriculture in the eighteenth and nineteenth century, with later forest regrowth after many people moved to the developing cites of eastern North America during the later nineteenth and early twentieth centuries. Figure 9.5b Summit of Green Mountain on Ascension Island. Prior to human intervention the largest common plants here were ferns (although the ferns visible in the photograph are introduced species). Now the summit is cloud forest dominated by bamboos, with trees covering much of the rest of the mountain. The sunny conditions in the photograph are deceptive; for much of the time the summit is enveloped in cloud that provides the water supply for this lush growth of vegetation on an otherwise arid island.

the early nineteenth century to one that today is best described as tropical cloud forest (Wilkinson, 2004a).

The archetypal pristine forest systems which have become an icon for much of the environmental movement are the tropical forests. The examples of both the seasonally dry and moist forests from Costa Rica and the Ascension Island cloud forest suggest that tropical forests may be able to recover quickly from serious human disturbance, in a similar way to the forests of eastern North America. Palaeoecological studies are now starting to show that many areas currently covered by tropical rainforest have had significant disturbance in the past. An early example of such studies was the Darian area of Panama which, being rich in endemic species, gives the impression of untouched tropical wilderness. However, Bush and Colinvaux (1994) used pollen and other remains from sediment cores to show a long history of human disturbance. Their data suggested significant forest clearance for agriculture that was abandoned after the Spanish conquest, so this apparently primeval tropical forest was largely 350 years old. As Paul Colinvaux (2007) later wrote, this seemed to show 'spectacular restoration of the forest once agriculture was stopped'—however, he suspected that this was most likely from the spread of trees from small patches of remaining forest, rather than the return of trees to an entirely treeless landscape. Similar evidence for widespread human disturbance of apparently 'virgin' rainforest has now been described from Africa and Asia, as well as South and Central America (Willis et al., 2004; Birks, 2019). Clearly rainforest systems may sometimes be much more resilient than has previously been assumed. It is interesting to compare this rapid recovery of recent forests with the evidence for rapid recovery of terrestrial biomass following the supposed impact event at the end of the Cretaceous (section 7.4).

Probably the most important factor in this fast recovery, and its associated carbon sequestration, is adequate rainfall. In the Ascension Island example, forest was able to colonize Green Mountain because of the trade winds blowing large amounts of moisture onto the summit (Ashmole and Ashmole, 2000; Wilkinson, 2004a). Modelling suggests that destruction of large areas of the Amazon forest could greatly reduce rainfall due to reduced transpiration and so lead to a climate unsuitable for tree growth

(Betts, 2004). If this is correct, a large biomass of vegetation in the Amazon is crucial to the continual existence of a climate able to support a large biomass of vegetation. Presumably the clearances of tropical forest now known from the archaeological record were not extreme enough to have such wider climatic effects.

One important result of thinking about the carbon cycle from a more geological perspective it that it draws attention to the crucial role of soil. In the long-term carbon cycle, soil is important because of its effects on silicate weathering. In the shorter cycle, it is the amount of organic matter contained in the soil that is important. How this will change with global warming is currently a large and important unknown in the behaviour of the Earth system (Cox et al., 2000; Lenton and Huntingford, 2003). The importance of soils in carbon sequestration is illustrated by a survey of Monk Wood, a well-studied deciduous woodland in southeast England, where it was estimated that approximately 129 tons of carbon per hectare were contained in the vegetation (including roots) and 335 tons of carbon per hectare in the soil (Patenaude et al., 2003). However, as is so often the case in ecology, there can be quite a bit of variation between sites, depending on site-specific details (Wilkinson, 2021a),

The extent to which land-use changes have affected soil carbon content is not well quantified, but it is potentially very important. For example, some cultivated soils are known to have lost over half their soil organic carbon through intensive agriculture (Lal, 2004). Therefore the management of soils, as well as aboveground biomass, is important in the context of increasing atmospheric carbon dioxide and climate change. The effect of past human land use on soils is also a potential complication for the ideas of Ruddiman on an early start to human effects on the global carbon cycle and climate. Many of the past modifications of the land will have reduced soil organic matter, which is consistent with Ruddiman's hypothesis. For example, Qiu et al. (2021) estimated that drainage and cultivation of peatlands between the years 850 and 2010 emitted some 72 gigatons of carbon, with 45% of this emitted before 1750. However, there are potentially important exceptions. For example, in Britain clearance of forest from wet, often upland, parts of the country played a role in triggering the formation of some blanket bogs (Moore, 1975, 1993; but see Gallego-Sala et al., 2016 for caveats). Such bog systems can sequester more carbon than the previous forest. However, the anaerobic nature of peat bogs creates an additional complication, as microbes within them produce methane (a greenhouse gas—see Box 8.2) as well as sequestering carbon. Globally, peatlands are very important in the short-term carbon cycle, as they contain around 600 gigatons of carbon. This equates to around 25% of the total global soil carbon stock (Loisel et al., 2021). As such, destruction of peatlands—be they temperate bogs in Britain or tropical peatlands such as those in South East Asia—is clearly important for the carbon dioxide content of the atmosphere. This is one of the reasons for the current interest in restoring damaged peatlands, as described for Chat Moss at the start of this chapter.

9.7 Carbon sequestration as a fundamental process

For any planet with carbon-based life, carbon sequestration appears unavoidable. Firstly, some organic matter is going to escape consumption by other organisms and become entombed in sediments. On any planet with oxygenic photosynthesis this is particularly important, as it potentially leads to increasing oxygen levels in the atmosphere. On Earth the silicate weathering cycle is a key process in the long-term carbon cycle. This is likely to be true of any Earth-like planet. As such, the evolution of terrestrial life (with its effects on weathering rates) is likely to increase rates of carbon sequestration. The obvious conclusion is that biologically controlled carbon sequestration is likely to be present on any planet with carbon-based life. However, it is important to realize that in general this aspect of life is an unavoidable by-product, not a specific adaptation. All else being equal, natural selection is likely to favour individuals that utilize carbon sources over ones that allow organic matter to escape to sediments. However, in the case of silicate weathering, adaptations that allowed access to mineral nutrients in rocks will have the additional effect of increasing weathering rates (Schwartzman and Volk, 1989).

Currently, on Earth, our industrialized societies play a major role in the carbon cycle and if Ruddiman (2005) is correct so have many pre-industrial societies. Since nothing can be said with confidence about the probability of life on other planets, it is clearly impossible to say much about the likelihood of species with human-like levels of agriculture and industry, although several scientists have attempted educated guesses (e.g. Shklovskii and Sagan, 1977; Davies, 1995). However, thinking about sources of energy for such a hypothetical species suggests they will be closely linked to carbon sequestration on their planet. An obvious early source of energy for a species that is going to develop agriculture and industry is burning biomass—it is hard to see how an ecology without oxygenic photosynthesis would be 'lively' enough to produce such a species, so the presence of an oxygen-rich atmosphere ideal for 'burning' is a good guess. Later, this biomass burning is likely to be followed by burning fossil fuels. As such, an intelligent species is likely to have profound effects on its planet's carbon cycle—before perhaps turning to technologies such as nuclear fission or fusion, or a range of more sustainable energy sources.

If this reliance on fossil fuels during industrial development is indeed crucial, then it allows interesting predictions about the history of industrial societies on any planet. As Fred Hoyle (1964) realized, this makes industrialization a 'one shot affair'. Consider a hypothetical major catastrophe overcoming industrial society on Earth, resetting the historical clock to a mediaeval level of technology. With easily accessible 'coal gone, oil gone, high grade metallic ores gone, no species however competent can make the long climb from primitive conditions, to high-level technology' (Hoyle, 1964, p 64). So, while we can say little sensible about the probability of an intelligent species with high technology evolving on another planet, we can predict with reasonable certainty that it will only happen once on each planet, or at least be separated in geological time by a period long enough for the regeneration of fossil fuel supplies outcropping at the planet's surface.

9.8 Overview

While most ecology textbooks only discuss the short-term carbon cycle, the role of life has been crucial in the geological long-term carbon cycle through processes such as silicate weathering. Interesting arguments have been put forward for the co-evolution of CO_2 levels and terrestrial plants—with adaptations to lower CO_2 levels allowing large leaves to evolve. It seems clear that without the effect of life (especially terrestrial vegetation) our planet would currently have a greenhouse-effect-controlled temperature that would likely rule out the survival of eukaryotic life. As such, it would appear reasonable to claim that carbon sequestration has a positive Gaian effect. However, this is possibly a local conclusion, suitable for Earth, but not able to be generalized to all other planets. If, as in the case of Earth, the output of the planet's star has increased over time, then reducing the carbon dioxide greenhouse may help to maintain life-friendly conditions, if the planet is not too far from the star. As the sun is thought to be a fairly typical 'main sequence' star (Narlikar, 1999), this conclusion will apply to many planets. However, if the planet is towards the outer limit of the habitable zone in its solar system (as defined by temperature affecting the availability of water), then lowering the greenhouse effect could lead to the cooling of the planet and reduce the life span of its biosphere. As such the fundamental ecological process of carbon sequestration could have a positive or negative Gaian effect depending on the particular situation on a given planet. More generally these ideas illustrate the importance of biomass as a key feature of global ecologies (see chapter 7)—the effect of vegetation (or phytoplankton) on carbon cycles is more directly linked to available biomass than species richness.

PART III

Emerging Systems

'There is a continued circulation of the matter of the surface of the globe from lifelessness to life, and from life back again to lifelessness.'

Huxley (1887, p 228).

CHAPTER 10

Nutrient cycling as an emergent property

10.1 The paradox of the goldfish

The goldfish *Carassius auratus* is native to Asia and parts of Eastern Europe (Maitland and Campbell, 1992; Lever, 1977). It has a long history of selective breeding in captivity, having been kept in China since at least the time of the Jin Dynasty (AD 265–420). The species was imported from China to Japan around AD 1502, while in Britain the first written record is from 1742 when Thomas Gray wrote his *Ode on the Death of a Favourite Cat, Drowned in a Tub of Gold Fishes* (Lever, 1977; Chen et al., 2020). In Britain it became traditional for children to keep goldfish as pets; in the past these were often kept in a glass bowl apparently devoid of other life forms—although in practice the water would have been rich in microbes. A hypothetical microbe-free goldfish bowl forms a useful counter thought experiment to the approach I have taken in this book (Figure 10.1). There is only one species, the fish, which survives because of a reliable input of fish food from outside the system. Many of the candidate fundamental processes described in Part II of this book are absent from this goldfish bowl system. A more realistic view of this thought experiment would be to imagine a planet inhabited by a single species, such as a microbe living in its ocean and being fed by a slow but constant input of extraterrestrial organic matter.

The approach I have taken in this book causes me to argue that the power of natural selection, and the important idea of ecological trade-offs, would lead to the evolution of a more diverse biota with multiple guilds utilizing each other's waste products—hence producing many of my fundamental processes and leading to the emergence of nutrient cycling. Until other planets with life are discovered, it seems impossible to test these ideas with real data. However, this does not necessarily mean that these ideas have to be based solely on intrinsic plausibility, as it is possible to approach these questions in a more rigorous manner: with both 'evolution in a test tube' studies and computer simulations.

10.2 Trade-offs and *in vitro* evolution

In chapter 4 I argued for a role for trade-offs in generating the conditions needed for biodiversity, suggesting that this idea should be given much more prominence in ecology than has historically been the case. While often not explicitly emphasized, this idea has been implicit in some ecological studies for many decades, a good example of this being the classic studies by Joseph Connell (1961) on barnacle distribution on the seashore at Millport in Scotland—see Grodwohl et al. (2018) for a science historian's take on this work and Wilkinson (2021a) for a more general description. Connell's results, described in many textbooks, implied that the fast-growing barnacle species *Balanus balanoides* (now called *Semibalabus balanoides*) appeared to trade-off its ability to survive the stresses of life high on the seashore against its ability to grow quickly, while the slow-growing *Chthamalus stellatus* (now considered to have been *C. montagui*) could survive in more stressful conditions. Philip Grime (1974, 2001) has similarly suggested a trade-off between

The Fundamental Processes in Ecology. Second Edition. David M. Wilkinson, Oxford University Press.
 DOI: 10.1093/oso/9780192884640.003.0010

Figure 10.1 The 'paradox of the goldfish', here on a slightly larger scale than a small glass bowl. In this case cultivated carp (which, as with goldfish, have a long history of domestication in China) are gathering to be fed in a pond in a park in Xi'an, central China. Although this pond will have more biodiversity than the 'goldfish bowl' described in the text it is unlikely that the nutrient cycling in this pond would support this biomass of fish without an external food source. Obviously in the case of a living planet such as Earth, cycling of nutrients, etc. has to be sufficient to support the long-term persistence of life, with only light energy coming from outside the system.

rapid growth and an ability to cope with stressful conditions in plants. These ideas have been around for some time; however, their general importance as an organizing principle in ecology has tended to be overlooked.

Studying evolution in action by experimental methods is often difficult because of the timescales involved. Quickly evolving systems, such as bacterial populations evolving *in vitro* or virtual organisms evolving *in silico*, can produce results on more realistic timescales, although experimental evolution studies were rather slow to realize the usefulness of microbes (Kawecki et al., 2012). A particularly interesting study, relevant to this book's principal thought experiment, was carried out by Rainey and Travisano (1998), using the common aerobic bacterium *Pseudomonas fluorescens*, which readily exhibits evolution under *in vitro* conditions. In their experiments they provided the bacteria with a range of ecological opportunities—afforded by spatial structure in the microcosm tubes. In a matter of days, replicate populations developed very similar morphological diversity, unless spatial structure was removed by shaking the tubes. The type of spatial structure available to the microbes is very similar to that available in a goldfish bowl: namely glass sides, surface film, and open water. Trade-offs appear very important in the evolution of this diversity; for example adaptations that suit colonization of the surface film are not good for colonizing the open water. As it is hard to envisage a planetary surface devoid of any spatial structure, these experiments provide some support for the idea that evolution would quickly lead to the breakdown of a single species goldfish bowl biosphere.

The experiments of Rainey and Travisano (1998) also illustrated some important points about frequency-dependent changes in an organism's fitness, which are very relevant to this book's main thesis. One of the important arguments against seeing a role for life in global regulation is the question, 'What about Genghis Khan species?'—i.e. species that effectively take over and destroy any community they reach (Pimm, 1991, p 190; see also Hamilton, 1995). The *Pseudomonas* microcosm experiments illustrate one possible answer to this question as these bacteria showed frequency dependence in the fitness advantages of various genotypes, so that they had advantages when rare that disappeared when they were common.

For example, one of the *P. fluorescens* morphs was named 'wrinkly-spreader' by Rainey and Travisano (1998); it formed a layer on the surface film of the water which gave it access to both oxygen in the air and nutrients from the microbial medium. However, as it becomes very common the wrinkly-spreader (which was reducing access to oxygen for the other morphs) became a victim of its own success; as the surface film mats increased in weight a point was reached where they could no longer support themselves on the surface and they sank. It is possible that any Genghis Khan species may often have the seeds of their own destruction hidden in their initial destructive successes, a point I return to in chapter 12 with discussions of 'sequential selection'.

These *in vitro* systems using *P. fluorescens*, or similar microbes, are a very informative way of studying evolutionary ecology (e.g. Buckling and Rainey, 2002; Kassen et al., 2004)—although there are obvious questions about how well the results scale up to the far larger spatial and temporal scales of planetary ecology. Such systems can potentially provide evolutionary insights into several of the fundamental processes described in this book.

10.3 Emergence of biogeochemical cycles

Clearly the development of trade-offs and adaptive radiations are highly relevant to the emergence of biogeochemical cycles, and the evolution of microbes *in vitro* is a very useful way of studying such things. The other main approach currently available is evolution *in silico* (Kawecki et al., 2012), as exemplified by artificial life studies of the evolution of nutrient cycling, for example the work of Keith Downing and colleagues around the start of the twenty-first century.

The GUILD model of Downing and Zvirinsky (1999) was particularly informative, as it simulated the evolutionary emergence of a web of interacting species whose combined effects can potentially control aspects of their chemical environment. This artificial life system uses a genetic algorithm (GA) approach. GAs are computer programs first developed during the mid-1960s that simulate the evolution of a population of genotypes, represented by a string of binary digits, with both sexual recombination and mutations of their genetic code (Holland, 1992). In the GUILD system the environment is represented by a number 'n' of different chemicals, while each organism's genes determine both the chemicals it feeds on and those it produces as waste products (organisms are not allowed to both feed on and excrete the same chemical). The organisms reproduce by division, with genetic variation coming from both mutations and gene exchange between organisms (sex *sensu lato*), and through the production and consumption of chemicals these organisms can create local ratios of nutrients that differ from the global ratios in the simulation. In these simulations the competition for resources leads to a diversity of guilds and the whole ensemble of guilds leads to the emergence of life-sustaining nutrient cycling.

Keith Downing (2002, 2004) also produced a second artificial life system with somewhat more realistic chemistry, which he called METAMIC. While GUILD essentially allowed all possible chemical reactions for the same metabolic cost, METAMIC used abstract chemistries that constrained the system to a smaller set of reactions and had more realistic relationships between energy, entropy, and biomass. The reduced number of allowable reactions in METAMIC restricts the number of metabolic options for the virtual organisms. This led to a wider range of possible outcomes for the whole system than in GUILD. The GUILD system showed a strong tendency to settle down to a stable state with regulated nutrient cycling. With METAMIC this happened in some runs of the simulation, while on other occasions regulation broke down. As Downing (2004, p 278) commented, 'if METAMIC has captured some essence of the real world, then one of the basic take-back-to-the-wild lessons from the computer runs is simply that Gaian homeostasis is neither tautological or impossible, but clearly contingent on a host of physicochemical, and quite possibly historical, factors'.

A more recent approach is the Flask model, which has also been used to study the evolution of nutrient cycling (Williams and Lenton 2007, 2008; Nicholson et al., 2017). In its simplest version this simulates a well-mixed container of liquid containing nutrients (that can enter and exit the flask) and is affected by an environmental variable (e.g. temperature, which microbial growth can also affect). Microbes are introduced that have simple 'genetics' that controls their interaction with both nutrients and the environment (and is capable of evolving). Readers with a background in microbiology will realize that this is effectively a computer model of a chemostat—a continuous culture device that can be used to experiment on microbial populations under laboratory conditions (Madigan et al., 2012). In more complex versions, multiple flasks can be linked together to create a more spatially heterogeneous environment (see Lenton et al., 2018 for a summary). In the context of this chapter the simple single flask model is of particular interest. As with Downing's models, nutrient cycling easily emerged in runs of the single flask model, but on occasions 'rebel' or 'Genghis Khan' organisms evolved that could crash the system. So again, regulation emerges but not in all cases (Williams and Lenton, 2007).

10.4 Cycling ratios and biotic plunder

The results of the *in vitro* and *in silico* studies described above suggest that we should expect biogeochemical cycles to appear as emergent properties of the presence of life on a planet, and that in many cases these cycles should have a homeostatic character that will tend to increase the probability of life surviving on a planet. Clearly it is those systems with some element of regulation that we would expect to survive for a long period of time, and so potentially to dominate the examples of living planets in the universe. This assumption may turn out to be wrong if life evolves from non-life very easily, which could lead to a large number of transient systems of short life expectancy that could outnumber the more long-lived stable 'Gaias'. However, at the other extreme, if it is the case that the initial origin of life is almost impossibly improbable then Earth may be the only example of such a system in the universe. If so, the observation that Earth-type life dominates the set of biologies in the universe is trivial, as the set only has one member and so it tells us little about the probabilities of Gaias developing, only about the unlikely nature of life itself.

Assuming a universe of relatively stable Gaias, what can we predict about the structure of their emergent cycles? A useful way of thinking about nutrient cycling is the concept of cycling ratios. This idea is due to Tyler Volk (1998) and usefully quantifies the way life on Earth cycles different nutrients. Consider calcium; life uses approximately 620 million tons a year, while around 550 million tons is lost each year from living systems (Volk, 1998). The ratio of these two numbers is $620/550 = 1.1$; contrast this with phosphorus, where Volk (1998) suggests that the comparable figures are $230/5 = 46$. The comparison of these cycling ratios 1.1 for Ca and 46 for P illustrates in a quantitative manner that phosphorus is cycled much more often through living systems before being lost to the non-living world. Evolution of nutrient cycles, of the type seen in Downing's simulations (described in section 10.3), should lead to higher cycling ratios for nutrients in short supply. In Downing's simulations his GUILD model produced cycling ratios of 30–40, while the (more realistic?) METAMIC simulation gave ones in the range 5–15. The results for phosphorus described above show that ratios of 46 are possible in the real world, even if METAMIC tended to evolve lower ratios (Downing, 2004).

The well-known capacity for unconstrained biological populations to grow exponentially suggests some potentially important generalizations about the concentration of nutrients in the environment. If all nutrients and other resources are common, then this should lead to unconstrained growth of the organisms, which will eventually use up the nutrients and so lead to an environment in which the nutrients are in short supply—a process that has been called 'biotic plunder' by Tyrell (2004). It is interesting that many cases of human-caused pollution involve problems created by artificially raised nutrient levels—such as the eutrophication of waterbodies in agricultural areas (e.g. Moss, 2001, 2012). Tyrrell (2004) mainly bases his arguments on plankton production in ocean systems, and viewed from an aeroplane most oceans and large lakes look blue, while viewed from underwater—with snorkel or SCUBA gear—visibility is often quite good. Both of these observations are true because of the low numbers of phytoplankton in most waters. This is mainly due to nutrient limitation, which can be demonstrated experimentally by adding additional nutrients. For example iron fertilization experiments in the Southern Ocean caused the sea to turn green and the responsible microbes to sequester increased amounts of CO_2 (Abraham et al., 2000; Watson et al., 2000). So, the presence of life in the oceans may regulate nutrients around low levels by rapidly using up any increased nutrients. Similar arguments have been applied to land, where one of the reasons the land is often green with vegetation is that plants are often poor food for animals, and so animal densities cannot build up to a point where they destroy the vegetation (Polis,1999; Wilkinson and Sherratt, 2016).

While Tyrell (2004) developed his ideas on biotic plunder based mainly on marine data, he also thought that a similar situation may exist in many terrestrial systems. A good example of this comes from Vitousek's (2004) long-term studies of the ecology of Hawai'i. He suggests that there is a plant/soil-microbial positive feedback system that tends to lower nutrient levels over time. He further suggests that high nutrient use efficiency by plants causes them to have lower nutrient levels in their tissues. This in turn leads to lower nutrient levels in the plant litter, leading to slower rates of decomposition and hence low nutrient availability, which in turn encourages high nutrient use efficiency in the plants. It is arguments of this type that connect these ideas to some approaches to the Green World question in terrestrial ecology.

How does all this fit with a Gaian view of planetary ecologies? Early versions of the Gaia hypothesis (e.g. Lovelock and Margulis, 1974) discussed a planetary 'homeostasis by and for the biosphere'. Such a phraseology tends to suggest regulation for optimum conditions for life, and such a definition looks unlikely to be correct in the context of the idea of biotic plunder. However, current versions of Gaia stress the 'habitable state' of the planet. For example, in an essay in *Nature* James Lovelock (2003, p 769) defined Gaia as the process by which 'organisms and their material environment evolve as a single coupled system, from which emerges the sustained self-regulation of climate and chemistry at a habitable state for whatever is the current biota'. Lenton et al. (2018) defined Gaia along similar lines as 'a coupled system of life on Earth and its abiotic environment [which] self-regulates in a habitable state, despite destabilizing influences such as a steadily brightening Sun, changing volcanic, metamorphic, and tectonic activity, and occasional massive meteorite impacts'. The idea of Gaian effect, used in this book (and in Wilkinson, 2003), comes from this emphasis on the maintenance of habitability which is potentially very different from the maintenance of optimum conditions. An interesting analogy with physiology is the well-known phenomenon that reduced calorie intake increases longevity in many animal species (Kirkwood and Austad, 2000). Clearly starvation is not optimal, but it can maximize longevity in a similar way to biotic plunder creating stable regulated conditions, although ones that are not optimum for biomass production. As Tyrrell (2004) pointed out, the idea of biotic plunder is compatible with the idea of life being involved in regulation and that this regulation is for a habitable state. However, this habitable state is not optimal for life if your definition of 'optimum' is based on high primary production or biomass, but it could be regulation for 'improved conditions' if this is defined as increasing the longevity of life on a planet—that is the Gaian effect. This appears sensible; while one can see in general terms how the presence of life as an important forcing factor in a system should make it more likely that the system stays within life-friendly conditions, it is not obvious why regulation for optimum conditions should be expected.

10.5 Overview

Arguments based on both *in vitro* and *in silico* models suggest that biogeochemical cycles will readily evolve on planets with life, along with many of the putative fundamental processes described in this book. Artificial life models illustrate the potential for these emergent cycling systems to have a positive Gaian effect. As Lenton and Latour (2018) put it, 'The participants in the recycling loop are no longer limited by what comes into their world, but rather by how efficiently they can recycle resources'. They went on to make the interesting applied point that currently humans are much less good at recycling than the rest of nature—we have much to learn from studying ecological cycles. Think, for example, of the level of efficiency shown by the cycling ratio for phosphorus described in section 10.4.

The well-known potential for exponential growth in unconstrained ecological systems suggests that these emergent systems will often regulate their environments around low nutrient states (biotic plunder) rather than at states that optimize productivity. This provides a context in which to understand why human-caused eutrophication is often a problem; if the natural state of many systems tends towards nutrient scarcity then it is not surprising that raising nutrient levels dramatically can sometimes have unwelcome effects. In the context of biotic plunder it makes sense to define Gaia in relation to prolonged habitability of a planet but not as a process that maximizes biological productivity at any one point in time.

CHAPTER 11

Historical contingency and the development of planetary ecosystems

11.1 Carus and the thunderbolt: chance and change in history

In AD 283 the Roman Emperor Carus was struck by lightning and killed; within a couple of years of this fatal accident both of his sons had been murdered and power had passed to a new Emperor whose approach to the administration of his empire was very different from Carus's (de la Bédoyère, 2002). A single unpredictable accident had changed the history of one of the most powerful groups of humans on the planet.

The narrative of human history is full of such pivotal chance events, and the seas surrounding my home in Britain have regularly provided illustrations of the important role of accident in history. In both 55 BC and 54 BC the Roman army made incursions into southern England from continental Europe, and on both occasions their logistics were thrown into confusion by storms in the English Channel. It was not until AD 43 that the armies of the Emperor Claudius successfully invaded much of Britain (Schama, 2000). In the summer of 1588 there was an even better illustration of the power of historical contingency in these same waters as a Spanish fleet made ready to invade England. The English ships under the command of Sir Francis Drake famously attacked this Armada off the south coast of England, but in truth Drake's efforts did only minimal damage to the Spanish, who sailed on to meet up with reinforcements in the Netherlands. Then 'the wind turned Protestant' and the resulting storm caused far more damage to the Catholic Armada than Drake had been able to achieve (Schama, 2000). A different weather pattern could have led to a very different outcome for England from 1588 onwards. Given the crucial global role this country played during the eighteenth and nineteenth centuries, this one storm may have affected the histories of countries from Canada to India, and beyond.

11.2 Historical contingency and ecology

The idea that similar accidents could be important in the broader history of life on Earth, as well as in the history of human societies, has been developed by several authors. One of the most eloquent advocates of the importance of historical contingency on geological timescales was the late Stephen Jay Gould (1990, 2002). He developed his arguments in a geological context based, in part, on the fossil faunas of the Burgess Shale (which are described in section 11.3). Although most palaeontologists now think he pushed these ideas too far, they are still good illustrations of the potential for historical contingency to have a major role in the history of a planetary ecosystem.

An interesting ecological example of the apparent role of chance is the history of the shrub *Rhododendron* in Britain and Ireland (Figure 11.1). *Rhododendron* is a large genus, with much of its species richness concentrated in South East Asia. The main species introduced to the British countryside was *R. ponticum*, an evergreen shrub about 2–8 m high that is native to parts of Spain and Portugal and also the area around the Black Sea and the eastern Mediterranean (Cross, 1975). The shrub was introduced into Britain as an ornamental garden plant around 1763 and was widely planted during the

The Fundamental Processes in Ecology. Second Edition. David M. Wilkinson, Oxford University Press.
 DOI: 10.1093/oso/9780192884640.003.0011

Figure 11.1 The spectacular flowers are one of the reasons that a wide range of *Rhododendron* species were introduced into British horticulture. Most Rhododendrons growing in the wild in Britain are, as in this example, mainly descended from *R. ponticum* introduced from Spain and Portugal; however, many also contain genes from a range of other (often North American) species. Cullen (2011) suggests 'there seems no possibility of recognising any kind of taxonomic units within the British stands' and suggests the name *Rhododendron* × *superponticum* for all individuals growing in Britain that show morphological evidence of hybridization (such as hairs on the underside of leaves), an approach now followed by the standard British flora (Stace, 2019). It's not yet clear how widespread *R.* × *superponticum* is in Britain (as it has only recent been formally described); however, some estimates are that up to 25% of the plants growing in the wild may be these hybrids, and this number could potentially increase with further study (Stace, 2019).

nineteenth century to provide cover for game birds. Today it is often considered a 'major alien environmental weed' (Dehnen-Schmutz et al., 2004).

As Chambers (1995) has pointed out, this history makes *R. ponticum* a particularly interesting example of the apparent role of historical contingency in ecology. Its natural distribution during the Holocene suggests a species limited to southern Europe and the Middle East—the obvious guess would be that its absence further north was due to the climate. However, the human introductions suggest that this shrub can thrive in much more northerly climes, being found in northern Scotland and at altitudes up to 600 m in the mountains of northwest England (Preston et al., 2002). We now know that this example is somewhat more complicated than was assumed in the 1990s when Frank Chambers (1995) used it as an example of the role of chance in dispersal, and it illustrates the way in which chance events often interact with other ecological processes. Molecular data make it clear that *Rhododendron* growing wild in Britain has, in many cases, acquired genes via hybridization from some North American species (which have also been planted in Britain) and these genes give increased tolerance to colder conditions. Such genes are commoner in *Rhododendron* sampled from the colder regions of Britain (Milne and Abbott, 2000). However, it seems highly likely that climate is not the only factor in the limited natural range of this species. This conclusion is supported by fossils of both *Rhododendron* pollen and seeds from sediments from a previous interglacial in Ireland—however, no fossil remains are known from mainland Britain (Godwin, 1975; Chambers, 1995). While this interglacial was warmer than the Holocene it still illustrates the shrub growing at much higher latitudes than its recent 'natural' distribution. It seems that a combination of chance dispersal (or lack of dispersal) and the acquisition of tolerance for colder conditions via hybridization provides the explanation for the absence of *Rhododendron* from the native vegetation of Holocene Britain and its recent success.

The potential importance of chance events, such as wind or bird dispersal of seeds, has already been discussed in chapter 5. If plant populations in part track climatic change with the aid of rare accidental dispersal events, then this clearly raises questions about the availability of such events over the short timescales predicted for human-caused 'global heating'. In addition, the habitat fragmentation caused by humans breaking up semi-natural vegetation with urban or intensive agricultural areas is likely to make it harder for plants to move through many current landscapes (e.g. see the models in Dyer, 1995). The chancy nature of dispersal also plays a role in the idea of assembly rules (Cox et al., 2016). This is the suggestion that the order in which species reach a location, such as an island, may matter. Two species could be equally able to colonize a site if they arrived there first, but are less likely to colonize successfully if a potential competitor has already arrived. The chance first arrival of species 'A' may rule out the successful colonization by species 'B'.

Additional examples of the role of chance in ecology are provided by models at the interface between population ecology and biogeography (e.g. O'Regan et al., 2002; Wilkinson and O'Regan, 2003). There is a general consensus in conservation biology that animals that are restricted to areas of a limited size (e.g. an island or a nature reserve) will only survive in the long term if this area is greater than a certain size—for a well-known example see Newmark (1987); but see Parks and Harcourt (2002) for an alternative and probably more realistic reinterpretation. One simple reason for this is that the smaller the resource base, the smaller the population that can be supported. The smaller the population, the more vulnerable it is to extinction through chance events (see section 6.4).

One of the difficulties with biogeographical studies is that no two cases are exactly the same, so that if you are working with species that are too large for artificially controlled experiments to be feasible, then every example is a special case. Consider a large organism, such as a big cat, living on an isolated area such as an island. Using a mixture of modern and historical data it is possible to conclude that a given population of that species is able to survive on an island of a given size (or alternatively we could conclude that the island is too small, if the species is known from fossils or historical sources but became extinct at a time when humans were probably not responsible). The obvious question of relevance to this chapter is, would such a population always survive or always go extinct on this island, or is an element of chance involved? One approach to answering this question would be to try to find other islands of the same size that have had populations of this species and ascertain if they managed to survive. However, there are bound to be other factors that vary between these islands (different climates, different competitor or prey species, etc.), so they are not really good replicates of the situation on the first island.

A theoretician's approach to this problem is to run replicate stochastic population models on a computer, so that you can be certain that each replicate does indeed have identical starting conditions—although as models they will necessarily be simplified caricatures of reality. The results of modelling of this type are shown in Figure 11.2. In these VORTEX models, a population of tigers *Panthera tigris* on an island the size of Bali always survived for a period in excess of 1,000 years, while the leopard *P. pardus* populations on 10% of the replicates of this virtual island became extinct. Figure 11.2 also shows the results for a population of medium-sized cats isolated in an area the size of Italy and largely based on the population biology of the modern jaguar *P. onca*. This was an attempt to study the ecology of the now extinct 'European jaguar' *P. gombaszoegensis* and its ability to survive glacial periods in southern European refugia. This is a point of some interest, as many temperate European organisms may have spent around 90% of the Quaternary in such limited ranges (refugia), as the glacial periods lasted much longer than the interglacials (see Figure 2 in Bradshaw, 1999). Although this putative Italian refugium was large (and most of the assumptions in the model were conservative, tending to increase the chance of survival), around 30% of the replicate populations became extinct over a 1,000-year period. These models illustrate the potentially large role for chance processes in the survival of populations of larger organisms. The situation may be very different for many free-living microbes, where huge population sizes may make extinction very unlikely (Fenchel, 2005), as does the ability of many taxa to form resistant 'spores' when conditions are unsuitable (Andrews, 2017). So chance may play a much smaller role in the ecology of the microbes that dominate many Earth system processes.

11.3 Historical contingency and the Earth system

Moving out of the Quaternary to the full expanse of geological time, there is huge scope for historical contingency. The most spectacular examples are the impacts of large meteorites on Earth (e.g. Grady et al., 1998). These can potentially cause extinctions, ranging from the loss of an endemic species of limited range size living in the area of impact to the destruction of all life on Earth, depending on the

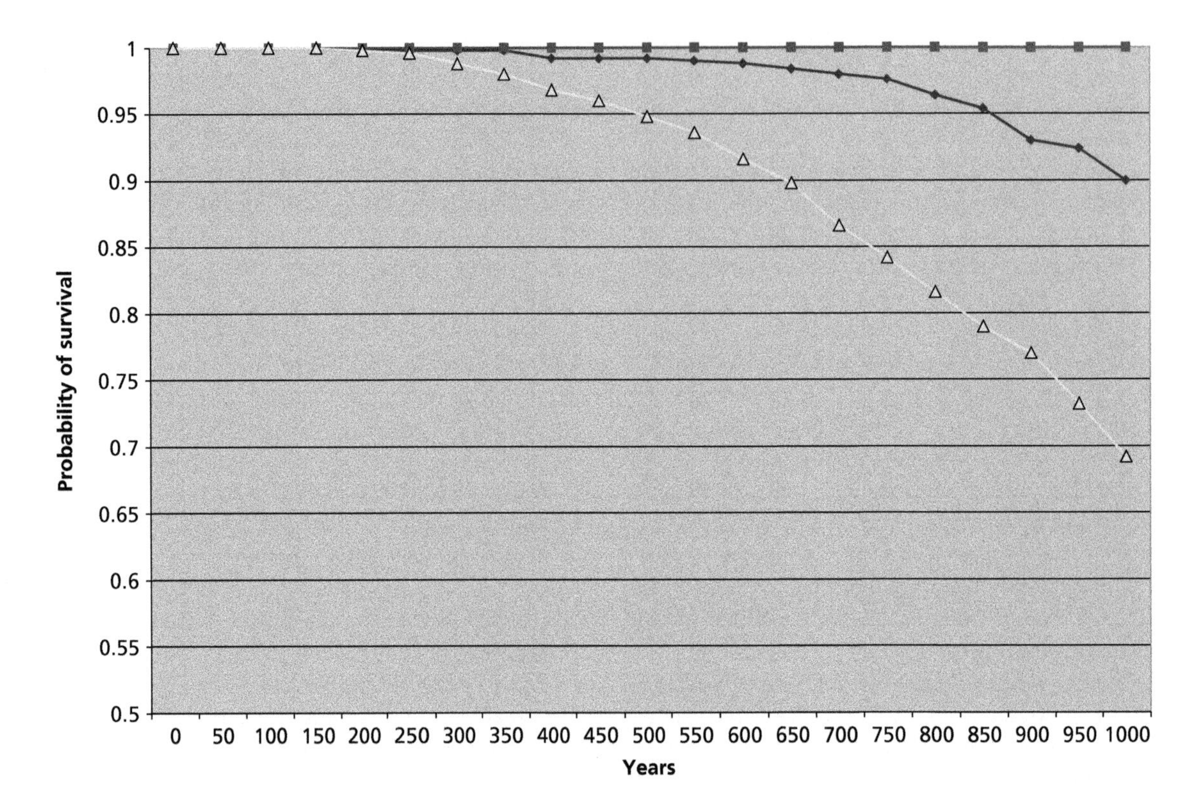

Figure 11.2 VORTEX models illustrating the role of chance in the survival of big cat populations. Squares (■) show results for replicate tiger *Panthera tigris* populations on an island the size of Bali, while diamonds (♦) show results for leopards *P. pardus* on an identical island. Triangles (▲) show the results for a cat of a similar size to the extinct *P. gombaszoegensis*, mainly based on the similar-sized jaguar *P. onca*, isolated in a glacial refugium the size of Italy (for background details on the palaeobiology of *P. gombaszoegensis* see O'Regan and Turner, 2004). These models were run using VORTEX version 8, with each curve being based on 500 iterations; for full details of the tiger and leopard models see Wilkinson and O'Regan (2003) and for *P. gombaszoegensis* see O'Regan et al., (2002). The key point illustrated by these results is the element of chance in the survival of two out of the three cat populations modelled. This arises directly out of the population dynamics in these models, as no environmental catastrophes or other additional variables (such as human hunting) were modelled. A ceiling carrying capacity, without an explicit description of density dependence, was used in all three models (c.f. the models in Figure 3.6; the legend to this figure also includes other details and caveats about VORTEX models). Lacy (2000) provided a general description of VORTEX.

size of the meteorite. Other potential triggers for catastrophic global extinction events include Snowball Earth and/or the oxygenation of the atmosphere. The survival of such mass extinctions may often be something of a lottery, with chance playing a role in which lineages survive to take part in subsequent adaptive radiations (Gould, 2002).

Probably the most well-known argument for the importance of historical contingency in palaeobiology is that associated with the Cambrian faunas recorded in the Burgess Shale. In part its prominence in the late-twentieth-century literature was due to Gould's (1990) accessible account of these faunas and what he viewed as their theoretical implications. The Cambrian is marked by an apparent 'explosion' in the diversity of the fossil record due to the evolution of shells and other hard body parts which more readily form fossils than does soft tissue. However, many of the animals in the sea at this time were still soft bodied and so less likely to fossilize. For a full picture of Cambrian biodiversity we need sites with unusual conditions (usually anoxic sediments) which allow soft tissue to fossilize. The most famous of these is the Burgess Shale in Canada (Conway Morris, 1989, 1998; Gould, 1990), although many other sites are now known from around the world (e.g. Conway Morris et al., 1987; Zhang et al.,

2001). These Burgess Shale-type faunas are characterized by great morphological diversity, although the exact extent of this is controversial (Conway Morris, 1998). During the 1980s it was suggested that there was probably a large element of chance in which of these lineages survived to give rise to the animal phyla we know today (Conway Morris, 1989; Gould, 1990); Conway Morris (1998) later became less happy with this highly contingent interpretation, as indeed are many other palaeontologists as further work has tended to reduce the extent of this morphological diversity and so the scope for chance to select which morphologies survive (Brasier, 2009).

A variety of explanations have been put forward to explain the 'explosion' in animal diversity during the Cambrian. These range from suggestions that this explosion is mainly an artefact of better fossil preservation to ecological mechanisms, such as a greater role for predators, or explanations based on developmental biology and genomic change (Gould, 1990; Conway Morris, 2000). In the context of this book some of the more interesting explanations are the ones based on Earth system processes. For example Lenton and Watson (2004) suggested that an early colonization of the land by microorganisms increased the rate of silicate weathering (section 9.2) which, via effects on atmospheric oxygen concentration, triggered the ecological changes behind the Cambrian explosion. A more conventional ecological explanation is the suggestion that the evolution of hard parts was driven by their potential to offer protection from predation (e.g. Conway Morris, 2000). In this context the identification of diffraction gratings on the surface of some Burgess Shale animal fossils (Parker, 1998) is very interesting, as colour now plays a prominent role in the ecology of avoiding predation, both in crypsis and warning colouration (Ruxton et al., 2018). It has also been suggested that increased bioturbation of sediments may have altered marine environments in ways that facilitated an increase in diversity (Herringshaw et al., 2017). Several of these mechanisms may have contributed to the real explanation, as it is often the case in ecology that there are several mechanisms involved and the key question is not which is correct but their relative contributions to the full explanation (Lawton, 1996). The relative balance between chance events (chance survival of a taxa, and/or impacts of large meteorites) and more predictable processes (predation, bioturbation) is currently an open question in long-term planetary ecology.

11.4 Overview

The ecology of a planet is influenced by historical processes; at any stage in its development the current conditions of life on a planet form the starting point from which new conditions develop. This means that over time an ever-increasing number of historical accidents will be incorporated into the system and so the role of past history will become increasingly important; this is an idea returned to in the final chapter (chapter 12) in the context of ideas of 'selection by survival'.

The idea of historical contingency is a simple one, hence the relative brevity of this chapter, and yet it is crucially important in understanding much of ecology. Consider oxygenic photosynthesis and the rise of oxygen levels in the atmosphere, which has rightly been described as a 'defining moment in Earth history' by Canfield (1999). This constrained the possible subsequent trajectories of ecological development on Earth, for example in relegating anaerobes to limited oxygen-free habitats. Such historical contingencies mean that the fundamental processes described in this book are necessary but not sufficient to explain the history of a biosphere; this is similar to the relationship between the ideas of *evolution* and *natural selection*. The opening sentence of the preface to one of the greatest theoretical books on natural selection states: 'Natural selection is not Evolution' (Fisher, 1930). While natural selection is necessary to explain evolution, it is not sufficient to understand how evolution has proceeded in a particular case as we need to know a range of historical details, many of which are effectively determined by chance events (Frank, 1998)—in the same way the history of England could have been different if a storm had not dispersed the Spanish Armada.

CHAPTER 12

From processes to systems

12.1 An Earth systems approach to ecology

My approach in this book has been to use an astrobiological thought experiment as a way of thinking about ecology on the scale of the whole Earth. This approach may be very important if ecology is to make major contributions to the mitigation of human-caused changes to the planetary system. Confronting problems such as global heating requires understanding of very large-scale systems, such as the atmosphere of the whole planet. There are several differences in emphasis between this approach to ecology and more conventional approaches exemplified by most university-level textbooks on the subject.

Firstly, microorganisms feature much more prominently in this book than they do in either ecology texts or most ecology journals. For example, for the first edition of this book I counted papers with a significant microbial content in a leading ecology journal and found that during 2005 only 5% of the papers in the journal *Oikos* had a strong microbial content. Sixteen years later in 2021 the figure was still only 5%, despite the large advances in the field caused by the rise of molecular approaches, such as next-generation sequencing (Clark et al., 2018). Other ways this book differs from conventional ecological thought are a tendency to give increased emphasis to the importance of biomass, rather than species richness, and an increased emphasis on parasites rather than predators. The relative lack of interest in parasites, compared to the huge literature on predation, was also identified as a problem with traditional ecology by Hamilton (1995) in his essay review of Pimm's (1991) book *The Balance of Nature?* This is still the case; for example predation is allocated twice the space of 'parasites and disease' in the widely used ecology textbook by Begon and Townsend (2021).

A succinct way of viewing the difference in the approach to ecology seen over much of the twentieth century and the more planetary perspective used in this book has been provided by Lewontin in multiple publications (e.g. 1983, 2002). He summarized the traditional evolutionary approach as a pair of differential equations:

$$dE/dt = f(E)$$

$$dO/dt = f(O, E)$$

where change in the environment (E) with time (t) is a function of various environmental variables—where 'E' is described in purely abiotic terms. The evolutionary change in organisms (O) over time is a function of both the abiotic environment and the biota. Lewontin suggested that the real situation is better represented by a pair of *coupled* differential equations:

$$dE/dt = f(O, E)$$

$$dO/dt = f(O, E)$$

so that changes in the abiotic environment are a function of both 'E' and 'O', as are changes in the organisms. At the larger spatial and temporal scales discussed in this book, these coupled differential equations are clearly a better way of thinking about the way things work, and are in part a return to the theoretical approach of Alfred Lotka in the 1920s (Kingsland, 1995). The key point is the two-way feedback between the biotic and abiotic elements. Consider one without the other and you are likely to be missing crucial parts of the system.

The Fundamental Processes in Ecology. Second Edition. David M. Wilkinson, Oxford University Press.
 DOI: 10.1093/oso/9780192884640.003.0012

Lewontin's evolutionary ideas have been developed in much greater detail under the title of 'niche construction' (e.g. Odling-Smee et al., 1996, 2003; Laland et al., 1999). This approach has focused on the evolutionary implications of the coupling between organism and environment and has developed a theoretical background derived from the mathematics of population genetics. A related idea that was developed during the 1990s is that of ecosystem engineering (Jones et al., 1994, 1997); this is a more ecological idea—compared to the evolutionary approach taken by niche construction. It points to the way in which some organisms greatly alter their environment with consequences for both themselves and other species—either by their physical presence (e.g. trees) or by their behaviour (e.g. beavers and humans) (Figure 12.1). Unlike niche construction, with its mathematical underpinning from population genetics, ecological engineering is grounded in natural history observation. Indeed the original paper in *Oikos* (Jones et al., 1994) describing the idea is highly unusual for a late twentieth-century theoretical paper in being all natural history and no mathematics!

Figure 12.1 Trees as ecosystem engineers. Riverine woodland in the south of Kruger National Park, South Africa—the main tree in the photograph is jackalberry *Diospyros mespilliformis* (the photograph was taken looking vertically upwards using a very wide-angle lens). Trees are good examples of autogenic ecosystem engineers, that is organisms that alter the environment by their physical presence. This can be contrasted with allogenic ecosystem engineers, such as beavers (Figure 7.1), which modify their environment through their actions—engineering in the same sense that it is used to describe human modifications of the environment (Jones et al., 1994). As can be seen in the photograph, the trees make the ground underneath their canopy shadier; they also affect other aspects of the environment for woodland floor species, for example reducing wind speed and often increasing humidity.

Both these approaches have emerged from within ecology and evolutionary biology over the last 30 years. However, the approach taken in this book has been heavily influenced by Gaia, an idea developed in the late 1960s and early 1970s (see Lovelock, 2000b for an autobiographical account and Dutreuil, 2018 for a science historian/philosopher's view). Gaia is clearly an ecological idea, although not developed by someone primarily trained as an ecologist. Indeed this approach was largely developed in ignorance of ecology during the 1970s, although Lovelock later came to see A.J. Lotka, Eugene Odum, and G.E. Hutchinson as forerunners of Gaia (Lovelock, 2000a). Hutchinson was an ecologist whose work nicely fits into the approach taken here; his wide academic interests included biogeochemistry and the application of cybernetic theory to ecological systems (Slobodkin and Slack, 1999; Slack, 2010). His work formed part of a mid-twentieth-century interest in processes at the ecosystem scale (see Odum, 1969 for a classic review), sometimes referred to as 'systems ecology'. Such large-scale studies seemed to go out of fashion during the 1980s and early 1990s but have become more common again. This revival of ecosystem-scale studies is probably due to interest and concern over global climate and related phenomena (Nobis and Wohlgemuth, 2004; Grace, 2019). In the context of this speculation it is interesting that Chambers and Brain (2002), in an extensive analysis of the published literature, suggested that 1988–1991 marked a major change in research publications, with an increase in papers on human-caused climate change and its potential effects.

12.2 The trouble with Gaia

Gaia implies not only strong coupling of life and the abiotic environment, as in Lewontin's coupled differential equations, but also regulation, contributing to the long-term persistence and stability of life on a planet such as Earth. The initial ideas on Gaia were developed, in part, using the analogy of Earth as a super organism (see section 2.2); this worried many biologists (e.g. Doolittle, 1981; Dawkins, 1982; Ehrlich, 1991; Williams, 1992a) because planetary regulation appears to be a more difficult phenomenon to explain than homeostasis in an organism. In most cases it is clearly in the interests of, for example, the cells in a multicellular organism to benefit other parts of that organism; however, this is not the case with Gaia. As Lenton et al. (2020) suggested, 'evolutionary biologists misread and misunderstood the idea of "regulation" and "homeostasis" as something akin to the common good'. Consider the production of DMS by some marine algae (section 7.3). This chemical has several functions in the Earth system: for example closing a loop in the sulphur cycle by transporting sulphur from the oceans back to the land, and also its role in cloud formation (Charlson et al., 1987). However, if these functions were costly then it should pay an individual cell not to contribute DMS to the Earth system. This is the well-known theoretical problem of 'cheats' in mutualisms (e.g. Wilkinson and Sherratt, 2001; Yu, 2001) but played out at a planetary scale. Around the late 1990s a partial solution to these difficulties was suggested: namely that Gaia must be built of by-products (e.g. Volk, 1998; Wilkinson, 1999b, 2004b; Lenton and Wilkinson, 2003). In the case described above the precursors to DMS would be produced for the cell's own needs, perhaps osmoregulation (Sunda et al., 2002), and their role in cloud formation and the sulphur cycle is 'accidental'. This immediately raises the question of why such by-products should have a positive, rather than a negative, Gaian effect.

From a humanities perspective, rather than using the language of evolutionary biology, the problem is that Gaia appears teleological. A teleological explanation is one that accounts for features—such as DMS production—by appealing to their contribution to the functioning of the whole system (Bogen, 1995). Some aspects of Gaia seem straightforward to explain using conventional natural selection. For example, nitrogen fixation by microorganisms is clearly an advantage to those organisms living in an environment low in reactive nitrogen. However, nitrogen fixation is energetically expensive and as the amount of reactive nitrogen increases in the environment the advantage of fixation becomes less (this can happen as fixed nitrogen leaks out of the fixing organisms into the wider environment). Therefore a system with nitrogen fixers and non-fixers may tend towards a stable state, where scarcity of available nitrogen causes an increase in nitrogen fixers, and so more nitrogen leaking into the environment, causing a decline in expensive nitrogen fixation as useable nitrogen becomes commoner in the environment (Lenton and Wilkinson, 2003). However, such examples seem to form the minority of postulated Gaian mechanisms, so it might appear as if a Gaian Earth system would need a designer if conventional natural selection cannot be used as the explanation for the working of large parts of the system. However, over the last decade a number of theoretical ideas have been developed which look promising as potential explanations for how Gaia could have 'evolved'; these are discussed in the next section.

12.3 Selection for Gaia?

In this section I briefly summarize some of these recent ideas that have mainly been developed since publication of the first edition of this book in 2006. For a more detailed account and more extensive references see Lenton et al. (2018). These ideas involve the potential for selective processes applying to the Earth system as a whole. Similar ideas date back to Arthur Tansley's 1935 paper introducing the idea of the ecosystem. In this he briefly speculated that, 'There is in fact a kind of natural selection of incipient systems, and those which can attain the most stable equilibrium survive the longest' (Tansley, 1935, p 300). A particularly noteworthy aspect of the recent development of ideas of selection acting on the whole system has been the involvement of Ford

Doolittle in their development. In a Gaian context he is known for one of the key early essays explaining why the idea looked very unlikely from an evolutionary perspective (Doolittle, 1981). However, in recent years he has been developing ideas on how communities of microorganisms could potentially evolve, ideas that he has also applied to Gaia (Doolittle, 2019; Inkpen and Doolittle, 2022). It is also the case that I have been a co-author on several of the studies discussed in this section—with all the obvious possibilities for bias in their description that this entails.

To give a feel for these approaches I describe two related ideas which have received recent attention: 'sequential selection' and 'selection by survival alone'. To understand these it is useful to be clear why natural selection operating on individual organisms (or individual genes) has a problem explaining Gaia (Wilkinson, 1999; Doolittle, 2019). As described in section 12.2, if selection is acting at the individual, or gene, level then any adaptation 'for the good of a wider group' is likely to be undermined by cheats following their own self-interest. In particular, natural selection is usually envisaged as acting on a population of competing and reproducing individuals—something that looks unlikely when considering the totality of life on Earth (for an alternative view see Cazzolla Gatti, 2018). As already described, this point was made repeatedly by evolutionary biologists during the 1980s and early 1990s. However, it is now clear that ideas of selection can be applied to clades rather than just individuals (Figure 12.2a,b). Here a clade could be anything from all the members of a particular species, to all plants, or even all life on Earth (as all extant organisms are thought to share a common ancestor). Selection at the species level, although controversial in the past, can meet the basic requirement for conventional ideas on natural selection, as species can reproduce (speciate) to form new species, although by definition these would not be new clades. However, this is not the case for genera, orders, or life itself (which clearly don't reproduce). However, as Doolittle (2017, 2019) points out, selection can still potentially act on such non-reproducing sets of organisms. Some clades will survive longer than others. This could be by chance (in which case it doesn't make sense to talk of clade *selection*); however, if their survival is influenced by properties of the clade then it makes sense to talk about clade-level selection. Clade selection has in the past been treated with some scepticism even by biologists with an interest in species-level selection (e.g. Gould, 2002). However, George Williams, one of the key originators of ideas about selfish genes, came to take clade selection seriously later in his career (Williams, 1992b), although he failed to see its relevance to his criticisms of Gaia (e.g. Williams, 1992a).

One of the key problems in trying to explain the longevity and stability of life on Earth as anything other than chance has been the difficulty in envisioning a role for evolutionary processes acting on a system that doesn't reproduce. In this context these ideas of clade selection look very promising. As mentioned, two ideas that take this approach are 'sequential selection' (Betts and Lenton, 2008; Lenton et al., 2018) and 'selection by survival alone' (Doolittle, 2014). To appreciate the idea of sequential selection, consider an evolutionary innovation that leads to some environmental effect. If this makes the system more stable, then that increases the probability that the system persists. However, if it destabilizes the system to the point of approaching the bounds of habitability (e.g. triggering a Snowball Earth event) then the resulting population crash may eliminate the organisms (clade) with the destabilizing trait (as well as lots of other organisms). However, if life survives this extreme event, it can effectively have another go at creating a more stable system. This does not imply that life is actively looking for stable systems; however, once a more stable system is established it is more likely to last (Lenton et al., 2018).

The related idea of 'selection by survival alone' focuses on persistence alone. The idea is that persistence of a system increases the likelihood of it acquiring further persistence-enhancing traits. Stable systems survive longer and so have an increased opportunity for acquiring further traits—including persistence-enhancing ones. The idea was initially developed as a graphical model (Doolittle, 2014); however, the basic ideas have now been demonstrated to be plausible in simple mathematical models (Nicholson et al., 2018b; Arthur and Nicholson, 2022). These are early attempts at addressing this

Figure 12.2 A botanical illustration of the idea of clade selection. A clade is a group of related organisms; indeed it is defined as containing 'all the descendants of a single ancestor (or ancestral species)'. A clade can have properties that don't apply to a single individual, such as number of individual members or its geographical distribution (Doolittle, 2019). Consider ginkgo *Ginkgo biloba* (Figure 12.2a); although many fossil species are known, this is now the only species in the order Ginkgoles (Christenhusz et al., 2018). Contrast this with sea mayweed *Tripleurospermum maritimum*, photographed in typical habitat on the Isle of Mull, Scotland (Figure 12.2b). This is a member of the daisy family (Asteraceae—which contains around 24,700 species) in the order Asterales. This order includes some 13.6% of all eudicot diversity—and the eudicots contain most of the diversity of flowering plants (Christenhusz et al., 2018). Therefore these two clades, with the same taxonomic rank, are very different as one contains just a single species that until recently was restricted to just a few locations in China, while the other has many species scattered around the world. The restricted diversity and distribution of Ginkgoles makes that clade at much greater risk of extinction than Asterales. Ginkgo is also interesting because, unlike so many other species, its interactions with humans have reduced its probability of extinction. Humans have planted it widely, both within China and around the world. For example, the photograph shows a young ginkgo plantation at Hanyangling Museum near Xi'an in central China, forming part of the attractions at the museum and associated tourist site at the tomb of Han Emperor Jingdi. However, much more important for long-term survival of the clade is the fact that ginkgo is now widely planted in cities across the globe as it is relatively unaffected by urban pollution—for example there are almost 60,000 ginkgo trees in the five boroughs of New York City, USA (Crane, 2019). Even so, restricted genetic diversity in Ginkgoles makes the clade more vulnerable to extinction than Asterales, with its many species and much more extensive genetic diversity. Genetic diversity is a feature of clades, not individual organisms, and so selection based on this diversity is at clade level.

idea quantitatively and there are still uncertainties about how best to formally model these ideas. For example, I think that in the mathematical models of Arthur and Nicholson (2022) their entropic Gaia model is probably closer to what Doolittle (2014) intended by selection by survival alone than their actual selection by survival alone model, as his graphical model implies the occurrence of mutations to the system which have the potential to improve survival. Arthur and Nicholson (2022) showed that in such a situation, 'increasing biomass, complexity and enhanced habitability over time is a statistically-likely feature of a co-evolving system'. Doolittle's emphasis on clades is problematic in one sense—it tends to overlook the non-living parts of the Earth system. However, it does give a central position to life, which seems appropriate (for example much of atmospheric chemistry is a product of life). Lenton et al. (2020) argue that the key difference between the Gaian approach and Earth systems science as widely understood is that Gaia gives emphasis to 'the centrality of Life'—whereas in the US government's National Aeronautics and Space Administration (NASA) conception of the Earth system, life is just one part amongst multiple component systems.

These ideas of the survival of systems are potentially applicable to systems other than the Earth system. For example, in the spirit of Tansley's 1935 suggestion of 'natural selection of incipient systems' in the paper that introduced the concept of ecosystems, they can potentially be applied to systems such as coral reefs and savannahs as well as human systems such as economies and beliefs (Lenton et al., 2021). These are mainly twenty-first-century ideas, new and yet to become widely established explanations. At our current state of knowledge the key point is not that they are indisputably correct (that's still uncertain), but that we now have a range of explanations that could potentially explain why a system such as Gaia might be expected to evolve on a planet with life. Other theoretical approaches are also available, such as the thermodynamic approaches described in chapter 2 (e.g. Rubin et al., 2020) and also ideas from learning theory (e.g. Watson and Szathmáry, 2016). This range of theoretical options means that one no longer needs to argue that the long-term persistence of life on Earth is just chance because there is no way such a system could evolve stability; however, that doesn't mean that chance doesn't play a role. The difficulty is that we only have one planet with life, which makes it hard to estimate the role of chance.

The palaeontologist Stephan Jay Gould (1990) had a useful thought experiment, albeit now a bit dated in its use of technological analogy. He suggested the idea of 'replaying the tape of life'—rewinding back through Earth history, deleting as you go, then allowing evolution to start again from some point in the past. Would you see a similar result or would chance events lead to a very different outcome this second time? One way to approach this question is to run computer simulations with virtual planets. An example of this approach is the study by Toby Tyrrell (2020), where he ran multiple versions of virtual planets with a simplified climate system and found that if the tape was replayed, with similar but slightly different starting conditions, then a range of outcomes was possible, with a planet sometimes keeping life-friendly conditions and sometimes not. So chance played a role in the long-term survival of life on a planet (see also Nicholson et al., 2018a). In his earlier book on Gaia, Tyrrell (2013, p 175) suggested, 'It is all about how much is luck and how much mechanism'. I think there is general agreement with this statement: luck must play a role (e.g. it seems reasonable that a more severe Snowball Earth event could have ended life on Earth, or that some extraterrestrial event could have sterilized the planet). The disagreements are in where the balance lies—I emphasize mechanism more than luck, while Tyrrell takes the opposite view (Wilkinson, 2015).

12.4 Conservation biology and the Earth system

This is not an applied ecology book; however, it is worth briefly considering how an Earth systems perspective affects approaches to nature conservation and related topics. As with academic ecology, conservation biology often focuses on entities such as species or habitats. Jepson and Canney (2003) have suggested that there is a 'recognition by scientists that their ability to represent nature in units ... creates the opportunity to integrate ecological theory with neoclassical economics', as an entity, such as a species, can be ascribed a value—this is something Jepson and Canny (in my view, correctly) considered not to be in the best interests of conservation. Species have often been used as what Simberloff (1998b) has referred to as 'umbrellas' in an attempt to conserve biological entities (Figure 12.3)—the idea being that a charismatic rare species (such as a big cat) can be used to justify the preservation of an area of land, which will also help preserve many other species (Simberloff, 1998b). The idea of nature reserves or parks is also entity-based.

This book has focused on processes rather than entities, taking a planetary-scale approach. What would this suggest for conservation? Haila (1999b) has argued that from this perspective the important things to safeguard are the processes themselves, not the named entities, be they tigers or rainforests. If the processes are protected then rainforest can recover and large predators can re-evolve (albeit on a long timescale). In this context, the arguments put forward in this book would suggest that we should pay more attention to biomass than to the numbers or variety of entities. Indeed important blocks of natural or semi-natural vegetation can be missing the kinds of entities that conservationists most often value. For example, Simberloff (1998b) pointed out that the 6,812,000-ha Tongass National Forest in Alaska did not have a single species that qualified for protection under the United States Endangered Species Act. Yet this National Forest would be contributing ecological services, as indeed do wholly unnatural vegetation types dominated by introduced species, although such vegetations are usually considered of no conservation interest (e.g. Wilkinson, 2004a). Here there is also a temptation to take an economic approach, assigning monetary values to these ecosystem services, but as with assigning an economic value to ecological entities, doing the same with ecosystem services (i.e. functions or processes) is fraught with problems, and likely often counterproductive (Silvertown, 2015).

It is possible that the great reduction in natural and semi-natural vegetation in many parts of the world could be compromising some of Earth's regulatory processes. Clearly this is an important research question. What is clear is that 'habitats' that are not normally considered of any interest by nature conservationists, such as ones rich in introduced species or plantation forestry, may still be playing a valuable role in the functioning of the

Figure 12.3 A male lion *Panthera leo* on a private game reserve in the Waterberg region of South Africa. The lion forms one of the 'big five' (elephant, both species of rhino, buffalo, lion, and leopard), the animals that tourists are often keenest to see in Africa. Conservation to favour these species can potentially benefit many other species, and savannah ecosystems in general (Lindsey et al., 2007). However, this may not always be the case; for example conservation may have sometimes caused very high elephant populations, which can have negative effects on some other species (Wilkinson et al., 2022). This reserve was also an example of rewilding; the land was previously farmed and had reverted to wildlife reserve because of the potential tourist income this provides.

Earth system, and that *such systems may be becoming increasingly important as the more natural systems decrease in area and biomass*. Such vegetations have traditionally not attracted the interest of most academic ecologists. It is time this changed, as we need to know much more about them and be able to say with more certainty how good they are as surrogates of natural vegetation. There is some evidence that this is beginning to happen, with some ecologists starting to argue that we should be thinking much harder about the ecology and management of these 'novel ecosystems' that are often rich in non-native species (Hobbs et al., 2006, 2013). These vegetations may even be increasingly important for the conservation of entities such as rare species, as climate change suggests that relying on nature reserves of reasonably pristine habitats is not a long-term solution, as these reserves will come to have the 'wrong' climate for the species they were originally set up to protect (Huntley, 1994; Pitelka et al., 1997).

However, amongst all these problems, there are some reasons for limited optimism. Semi-natural and, in some cases, wholly artificial vegetation can sometimes rapidly develop and carry out various Earth system processes (section 9.6), and some more low-intensity human land uses produce habitats that are greatly valued by many conservationists (Figure 12.4a,b). To take an optimistic view, human introductions have greatly *increased* the number of plant species found in many parts of the world (Thomas, 2017). While the Earth systems approach is currently in its infancy, it may eventually lead to a very different emphasis for the conservation movement. A focus on entities may be no longer practical in many cases in our rapidly changing world; a preservation of the processes that will eventually allow recovery may be the best we can achieve. However, as has been illustrated by multiple examples in this book, biodiversity still matters, as it feeds into many ecological processes. As Hambler and Canney (2013) point out, 'Ultimately, Gaia could prove one of the most compelling reasons for conservation'.

12.5 Concluding remarks

A systems view of nature, replete with feedbacks, is a remarkably recent way of viewing the world. In European thought the dominant technological metaphor of many centuries was clockwork—something that should run perfectly, without the need for feedback once it had been set going. During the eighteenth century, especially in Britain, it became increasingly common to find ideas of 'checks and balances' and self-regulation, especially in political and economic theory (Mayr, 1986; Henry, 2002). Technology utilizing feedback had been known from at least the first century AD; however, these were 'toys' rather than devices of practical use. Feedback starts to become technologically important in eighteenth-century England, first for regulating windmills and later for regulating steam engines—for example the famous Boulton–Watt centrifugal governor of 1788 (Mayr, 1986). These ideas of feedback and checks and balances moved from economics and technology to medicine during the nineteenth century, most importantly in the work of Claude Bernard in France on the idea that would later become known as homeostasis (Porter, 1997). Almost 100 years later, during the Second World War, physiologists studying the aiming of anti-aircraft guns came to appreciate the crucial importance of feedback processes in the interaction of muscles and brain, an insight they applied to post-war medical research (Miller, 1978). Note, however, that all these examples come from a European perspective. Xiaolong Song et al. (2022) suggest that some aspects of a Gaian approach resonate with various philosophical views from ancient China—especially Taoism. The traditional Chinese approach to what would later be called 'science' put more emphasis on correlation rather than causality, seeing the universe as something like a vast organism (Needham, 1956; Cohen, 2015). There is also a history of Gaian-like ideas in the Russian tradition, especially associated with Vladimir Vernadsky's early twentieth-century ideas on the geological role of living organisms (Lekevičius, 2006). However, these non-western Gaian linkages mainly apply to seeing life embedded in the wider world, not specifically in the importance of feedbacks and regulation.

In the context of Gaia it is probably relevant that James Lovelock spent the first part of his scientific career working in medical research—for example playing a crucial role in elucidating the mechanism by which cells are damaged by freezing, a result

Figure 12.4 Examples of human-created habitats of conservation interest because of their biodiversity. Traditionally managed hay meadows are a good example of unnatural systems that are high in plant biodiversity (Wilkinson, 2021a). Disturbance of cutting prevents a few plant species from dominating the site, as do relatively low nutrient levels due to lack of modern chemical fertilizers. Our modifications to the world do not have to cause mass extinctions; however, such low-intensity use probably requires a much lower human population size than currently found on Earth, if it is to be widespread. Figure 12.4a Forest Meadow at Råshult (birthplace of Linnaeus) in southern Sweden. Pollen analysis has shown that this meadow system was created, from woodland, about 900 years ago (Lindbladh and Bradshaw, 1995). It is still managed by traditional methods—note hay drying in foreground of picture—with the aid of subsidies, as this is no longer an economic way to farm in Sweden. It is rich in plant species, as is appropriate for the birthplace of one of the greatest names in taxonomy! Figure 12.4b Traditional, species-rich, alpine meadow in the Italian Dolomites. Because of the steep terrain, it is still managed by traditional methods (the mechanical harvester in the picture being little larger than a garden lawn mowing machine).

with huge implications for reproductive medicine and the storage of eggs, sperm, and embryos (Pegg, 2002). Ideas of feedback already existed in the environmental sciences, such as the 'systems ecology' of Hutchinson and the Odum brothers during the mid-twentieth century; Hutchinson was influenced in his approach by the work of Vernadsky in Russia (Hutchinson, 1970; Slack, 2010). However, Gaia was marked out by a particular emphasis on feedback and regulation, which was unusual in the environmental sciences in the 1970s; indeed the interest in ecosystem processes in ecology was about to suffer a decline. As John Lawton (2001) has pointed out in an editorial on Earth systems science in *Science*, 'James Lovelock's penetrating insights that a planet with abundant life will have an atmosphere shifted into extreme thermodynamic disequilibrium, and that the Earth is habitable because of complex linkages and feedbacks between atmosphere, oceans, land and biosphere were major stepping-stones in the emergence of this new science'. The term 'Earth systems science' was first used at NASA, with their Earth systems science committee established in 1983 (Steffen et al., 2020).

The Earth system is obviously very complex, and humans are currently altering it in many ways. A Gaian approach tries to organize a lot of information in a way that allows one to ask interesting and hopefully useful questions. In particular, it forces us to think hard about feedbacks and gives life in general and microbes in particular the central place they deserve in the working of the planet. In the context of the processes described in this book the ecologically most important group is the prokaryotes, followed by the single-celled eukaryotes and then the fungi and plants (perhaps of equal importance). The least important group is the animals. The striking thing about this ranking is it is almost exactly opposite to the amount of attention given these groups by ecologists!

In the terminology of Vepsäläinen and Spence (2000) Gaia is not really a scientific theory (a relatively constrained statement about the world susceptible to a single test which could find it wanting) but an 'explanatory framework'. Such frameworks are 'road maps to solutions, rather than solutions themselves . . . so that an investigator can pick and choose what is required to effectively understand a specific event or situation' (Vepsäläinen and Spence, 2000, p 211). It is now clear to many scientists that it is impossible to understand a planet such as Earth without considering multiple feedbacks between life and the abiotic environment. It is also clear to many that Earth exhibits a certain amount of regulatory behaviour, in which life is intimately involved. Key questions for the future are about the strength of these regulatory processes and the mechanisms underlying them, and how the fundamental ecological processes contribute to the working of the Earth system.

Glossary

This book considers ecology in the broader context of Earth systems science; as such it ranges across a number of scientific disciplines. This created problems in the use of various technical terms, which come not only from ecology but also from a range of other disciplines such as astronomy, climatology, geology, chemistry, and physics. In an attempt to reduce this problem the glossary defines many of the technical terms used in this book; in addition I have included some comments that would have broken the flow of my principal arguments if inserted in the main text. In composing the definitions I have assumed that anyone looking up a term probably comes from an area of science different from the one where the term is in common currency. As such the glossary attempts to give the basic idea behind the term, rather than aiming for a fully rigorous formal definition. In addition where I cite a text for further information I have, in most cases, deliberately chosen less technical texts (usually at a level accessible to someone starting to study that subject at university). Two excellent textbooks that attempt to take an Earth systems approach to ecology and related areas of science are Kump et al. (2010) and Boenigk et al. (2015). *Ecology* by Ricklefs and Relyea (2014) is a nice introductory conventional ecology textbook, and my own book *Ecology and Natural History* (Wilkinson, 2021a) introduces key ecological ideas in a relatively non-technical way mainly using examples from British natural history. A good dictionary of biology is that of Hale et al. (2005), while Calow (1999) provides a useful dictionary of ecology. The glossary is followed by a brief summary of the geological timescale on Earth.

Abiotic environment: Traditionally ecology textbooks split the environment into 'biotic' (biological aspects of the environment) and 'abiotic' (physical aspects). Abiotic would include factors such as aspects of climate, soil pH, and oxygen concentration of the atmosphere; however, the approach taken in this book makes it clear that this classification is often unhelpful as most of the traditional abiotic aspects of the environment are affected (perhaps even regulated?) by biology.

Adaptive radiation: The evolution of a wide range of related forms from an ancestral species.

Albedo: Also called 'reflection coefficient', the reflectivity of an object. For example the albedo of fresh snow is around 0.95 while a flat calm ocean can be as low as 0.20 (see Pielou, 2001).

Anthropic principle: The idea that only some types of universe could possibly support observers (astronomers, etc.), so it is not surprising to find that we live in such a universe. This has been an important but also difficult and controversial idea in astronomy (see Smolin, 1997 for a sceptical but relatively non-technical discussion).

Anthropocene: Suggested name for the very recent geological past (in most definitions the last 200–250 years) in which humans have been altering the environment to such a large extent that they have become a major geological force. As Tickell (2011) pointed out, 'The idea that humans could so transform the land surface, seas and atmosphere of the Earth to establish a new geological epoch in their own name is startling in itself, and would have amazed earlier generations'. See Crutzen (2002) for a short essay on the idea, written by the scientist responsible for popularizing the idea of the Anthropocene.

Archaea: Originally called 'archaebacteria'. In the late 1970s Carl Woese and George Fox showed that the 'bacteria' actually consisted of two very different groups of organisms—the archaea and the 'true' bacteria which are now usually considered to form two different domains of life, with 'domain' being a higher classification than the more familiar 'kingdom'. (See Eme and Doolittle, 2015.)

Archean: The geological eon covering the period of time from 2,500 until approximately 4,000 million years ago (also spelt archaean). The start of the Archean is not well defined; a logical place to draw the lower boundary would be at the first appearance of life. However, as described in chapter 1, the dating of this event is highly controversial.

Autotroph: An organism that can use non-organic sources of energy, the most well-known examples being green plants using solar energy.

Biomass: The total mass of living material. For example the biomass of a forest would be the combined mass of every organism living in that habitat.

Biome: An area of the world with similar ecologies. Examples include 'tropical rainforest' and 'temperate deciduous forest'.

Biosphere: This term has been used in several different ways. Some authors use it to mean the 'totality of living things residing on the Earth', others use it to mean 'the space occupied by living things', while others use it to refer to 'life and life support systems' (Huggett, 1999). In this book the second definition is used; that is the biosphere is that part of the planet occupied by life. The third 'systems' definition has much in common with Gaia (*q.v.*). For an English translation of the classic text entitled *The Biosphere* first published in 1926 see Vernadsky (1998).

Boreal forest: The vast forests of coniferous trees forming a zone between tundra and deciduous forest in the northern hemisphere.

Carrying capacity: The maximum number of individuals that an area can support, such as the carrying capacity of a particular bird species on an island. It is often represented as 'K' in the logistic equation (*q.v.*) (see chapter 8 in Wilkinson, 2021a).

Chloroplast: A structure within plant cells that contains the photosynthetic pigments. They evolved from free-living photosynthetic bacteria. The term chloroplast was coined by Andreas Schimper in 1883; he also speculated that these chloroplasts may have had a symbiotic origin. However, he didn't develop these ideas further, but went on to be a key figure in the early history of plant ecology (see Sapp, 1994 and Quammen, 2018).

Clade: A group containing an ancestor and all its descendants. The largest clade is all of life on Earth, as all existing life is thought to share a common ancestor.

Cloud condensation nuclei: Substances around which water droplets condense during cloud formation.

Coccolithales: Planktonic protists ('protozoa') in the phylum Haptomonoda; they are covered in plates of calcium carbonate, which makes them important in the marine carbon cycle (see Margulis and Chapman, 2009 and Boenigk et al., 2015).

Cosmic rays: Elementary particles (e.g. protons, electrons) travelling through space.

Cosmopolitan: An organism that is widely distributed; usually applied to species found on most continents, or in most oceans if marine.

Cyanobacteria: Oxygenic photosynthetic bacteria; these were often referred to as blue-green bacteria (or 'algae') in the past (see Margulis and Chapman, 2009 or for details of their ecology Whitton, 2012).

Cybernetics: The science of control systems, often used as a synonym for systems theory.

Detritivore: An organism that gains its energy from dead organic matter (detritus).

Diagenesis: Modifications to sediments, etc. which occur as they are changed into rock.

Diatoms: Aquatic single-celled eukaryotes with silica-rich shells (see Margulis and Chapman, 2009 and Boenigk et al., 2015).

Diffraction grating: A collection of many microscopic grooves which reflects light in such a way that it is broken up into its constituent colours (see Parker, 2005).

Dinoflagellata: A group of planktonic protists ('protozoa'). At least 2,500 species are known—both photosynthetic and heterotrophic. Also referred to as Dinophyta (see Margulis and Chapman, 2009 and Boenigk et al., 2015).

Ecosystem engineering: It has long been known that organisms modify their environment; however, during the 1980s Clive Jones and colleagues formalized this into the concept of 'ecological engineers'. These organisms have a particular effect on their environment, either by their physical structure (e.g. trees) or by their behaviour (e.g. beavers). Extinction of such species may have major effects on the environment (see Figure 12.1).

Ecosystem services: Services of use to humans and other organisms provided by ecosystems, e.g. oxygen, food, 'clean' water.

Emergence: The formation of global patterns from local interactions; density-dependent regulation is the most well-known ecological example (see Stewart, 1998). This concept was first named by the biologist G.H. Lewes (now better known as the partner of the novelist George Eliot) in the nineteenth century, but attracted relatively little interest until the late twentieth century (Ziman, 2003).

Encyst: Entering into a protective encapsulated form (often a dormant or 'resting' stage) usually in response to extreme environmental conditions.

Entropy: A measure of disorder of a system or substance, it can also be considered to be energy in a useless form (see Box 2.1 and Pielou, 2001).

Eukaryotes: One of the two great structural divisions of life on Earth; the other is the Prokaryotes (*q.v.*). They have a more complex cell structure than Prokaryotes, usually with organelles such as mitochondria (see Margulis and Chapman, 2009).

Food chain: A sequence describing feeding relationships. For example Plant → Caterpillar → Tit (Chickadee) → Hawk, where the arrow shows feeding relationships. So in this example caterpillars eat plants but tits eat caterpillars (see chapter 3 in Colinvaux, 1980).

Food web: An interconnected network of food chains (*q.v.*) describing the feeding relationships for a whole ecological community, such as a woodland or lake.

Free energy: Energy capable of doing useful work (see Pielou, 2001).

Gaia: The idea, suggested by James Lovelock, that organisms and their material environment evolve as a single coupled system, from which emerges the sustained self-regulation of climate and chemistry at a habitable state for whatever is the current biota.

Geophysiology: A term sometimes used as a synonym for Gaia (*q.v.*).

Gray: A unit of absorbed radiation dose (1 Gy = 100 rad, where 1 rad = 100 ergs/g).

Group selection: Natural selection between groups (rather than individuals); in most cases this is a much less powerful force than individual-level selection. There is considerable semantic confusion in the literature over definitions of group and kin (*q.v.*) selection.

Guild: A group of species that all make their living in a similar way. Named for a perceived analogy to medieval guilds of tradesmen; however, the guilds of medieval England were rather more complex institutions than this analogy suggests.

Heat death: A state of uniform temperature (e.g. of a planet or the whole universe) where entropy (*q.v.*) is maximal.

Holarctic: A biogeographical region composed of the Palaearctic (mid- and high-latitude Eurasia, including North Africa) and Nearctic (America, north of tropical Mexico).

Holocene: The last 11,650 years since the end of the last glaciation. The last part of the Quaternary. (See the geological timescale at the end of this glossary.)

Homeostasis: Maintenance of a constant internal environment by an organism (such as human body temperature).

Host: The organism upon which a parasite (*q.v.*) lives and, usually, feeds.

Hyphae: Microscopic filaments that form the basic structure of a fungus (see Sheldrake, 2020).

Information: In the technical sense of Shannon (named after Claude Shannon), information can be considered the reciprocal of entropy (*q.v.*).

Kin selection: A special case of group selection (*q.v.*) where individuals are closely related—and so will have many genes in common.

Littoral: Living at the margin of a water body, e.g. at the edge of a lake or in the intertidal zone of a sea.

Logistic equation: The simplest equation that relates the rate of growth of a population to that population's density, so that

$$dN/dT = rN\,(K - N/K)$$

where N is population size, r is population growth rate, and K is carrying capacity (*q.v.*).

Lotka–Volterra models: These model the competition between species based on the logistic equation (*q.v.*), so that the rate of population growth of species 1 when in competition with species 2 can be written as:

$$dN_1/dT = r_1N_1\,((K_1 - N_1 - \alpha N_2)\,/K_1)$$

where α is a competition coefficient and is effectively a way of expressing species 2 in units of species 1 (for example species 2 may be larger and require more resources than species 1). d, N, T, r, and K are as in the logistic equation.

Macroparasites: Multicellular parasites, such as nematodes and lice. They reproduce much more slowly than most microparasites (*q.v.*).

Magnetosphere: The volume of space around an astronomical body, such as a planet, where its magnetic field plays a prominent role.

Main sequence: This describes a group of stars on the Hertzsprung–Russell diagram (a plot of a star's surface temperature against absolute luminosity). Our sun is a main sequence star (see Narlikar, 1999 or any astronomy textbook).

Mass ratio hypothesis: The idea that the extent to which a plant species affects ecosystem function shows a strong positive correlation with its biomass. This idea has been implicit in many discussions of plant ecology but was formalized by Philip Grime in the late 1990s.

Mesocosm: A controlled experimental system, ideally closed to external influences. A small one would be a microcosm while a very large one would be a macrocosm; however, there are no generally accepted boundaries for when a microcosm becomes a mesocosm, etc.

Metamorphic: Rocks that have been altered by extreme heat and/or pressure.

Metazoa: Animals excluding 'protozoa'-like organisms that traditionally were often considered to be animals or at least to fall within the area of interest of zoology. More formally, all organisms developing from a blastular embryo.

Microparasites: Small parasitic organisms such as viruses, bacteria, and protists. Often capable of extreme multiplication within the host (*q.v.*) as they have very short generation times. *C.f.* macroparasites.

Molecular clock: Using the rate of changes in biological molecules (e.g. genetic material or proteins) to estimate the time since two organisms diverged—such as the date of the last common ancestor between humans and other apes.

Mutualism: A mutually beneficial relationship, usually used to describe the relationship between different species. See also symbiosis (*q.v.*).

Mycorrhiza: An association between plant roots and fungi that is normally beneficial to both species (see chapter 7 in Wilkinson, 2021a).

Niche construction: The process by which organisms modify their own or other organisms' environments. This approach was developed during the 1990s by John Odling-Smee, Kevin Laland, and Marcus Feldman. It emphasizes the evolutionary effects of this process and has its theoretical origins in ideas from population genetics (for a non-technical account see Jones, 2005).

Occult precipitation: Deposition of water directly from clouds, e.g. by deposition onto tree branches, etc. Also called 'cloudwater deposition' and 'horizontal deposition'.

Organic chemistry: The chemistry of carbon compounds. Note this does not imply that they are of biological origin—for example many organic chemicals have now been identified in space; however, the majority of astronomers believe all of these are of abiological origin.

Palaeosol: An ancient soil, which may have undergone lithification (i.e. been turned to rock).

Palaeozoic: A geological era covering the period of time between 542 and 251 million years ago.

Panspermia: Literally 'seeds everywhere', the idea that life arrives on a planet from elsewhere in the universe. A variant is the concept of directed panspermia (an idea particularly associated with Francis Crick and Leslie Orgel) where microbial life is deliberately spread through space by an intelligent species (Crick, 1982).

Parasite: An organism that lives in or on another organism and in so doing obtains resources at its host's (*q.v.*) expense. It is remarkably difficult to produce a rigorous definition of parasite which successfully covers the wide range of examples seen in nature (Box 3.2). See also microparasites and macroparasites (*q.v.*).

Parasitoid: An organism that lays its eggs inside another organism and the growth of the parasitoid ultimately kills the host. Classic examples are some species of wasps and flies.

Partial pressure: In a mixture of different gases partial pressure is the pressure that a gas would exert if it occupied that volume by itself.

Peat: The highly organic remains of plants and animals accumulating under waterlogged conditions (see Figure 7.4 and Rydin and Jeglum, 2013).

Pelagic: Living in open water.

Phagocytosis: The engulfing of solid particles by a cell by the budding off of a vesicle from the cell membrane.

Phanerozoic: The geological eon covering the most recent 541 million years of Earth's history (from the start of the Cambrian until the present). For its subdivision into geological periods see the table at the end of this glossary.

Photorespiration: Competition between O_2 and CO_2 for sites on the enzyme rubisco; this process can reduce the efficiency of photosynthesis (see Morton, 2007).

Photosystem I and II: Two light-capturing molecular systems involved in the 'light reaction' of photosynthesis (see Morton, 2007).

Predator: An organism that consumes another, killing it in the process. It is remarkably difficult to produce a rigorous definition of predator that successfully distinguishes it from parasite (*q.v.*) (see Box 3.2).

Primary production: The rate of production of biomass (*q.v.*) by autotrophs (*q.v.*) (see chapter 11 in Pielou, 2001).

Primary succession: The process by which life colonizes bare ground (*q.v.* succession) (see chapter 9 in Wilkinson, 2021a).

Prokaryotes: Single-celled organisms lacking a nucleus and other organelles, bacteria *sensu lato.* One of the two great structural divisions of life on Earth; the other is the eukaryotes (*q.v.*). However, the status of prokaryotes as a group is controversial since the splitting of archaea from bacteria in the 1970s.

Proterozoic: The geological eon from 541 to 2,500 million years ago (the late 'Precambrian').

Pseudopodia: Temporary projections of cellular material in some protozoan cells, used in feeding or locomotion.

Quadrat: An area, often square, marked out on the ground to help record the distribution or abundance of species.

Reactive oxygen species: Oxygen-containing molecules with an unpaired electron (free radicals).

Reduction: In a chemical context a reaction in which a molecule has oxygen removed or hydrogen or electrons added; also the addition of electrons to an atom. In philosophy the act of explaining concepts or processes at one level in the terms of those from another level, usually seen as more basic. The often-heard statement that 'all biology is just chemistry' is an example of a reductive approach to biology—and ignores crucial ideas such as natural selection or adaptation which are not present in chemistry.

Second law of thermodynamics: At its simplest this states that heat flows from a hot body to a cold one; more formally that the free energy (*q.v.*) of a system decreases in any spontaneous change (see Pielou, 2001 and chapter 2).

Snowball Earth: The idea, particularly associated with Paul Hoffman, that at various times in the geological past Earth has suffered major ice ages which led to most of the planet freezing. People who argue for less extreme events often write about a 'Slushball Earth' (see Lenton and Watson, 2011).

Stefan–Boltzmann constant: These two physicists showed that the energy given off by a perfect or 'black body' radiator is $B = \sigma T^4$, where σ is the Stefan–Boltzmann constant which is 5.67×10^{-8} Wm^{-2} K^{-1} and T is temperature. In the context of the Daisyworld model it is relevant that within the range of temperatures typical of Earth's surface, soil and vegetation approximate to black body radiators (Monteith and Unsworth, 1990).

Stomata: Opening in the surface of a plant (often in the leaves) which allows gaseous exchange between the plant's interior and the surrounding atmosphere.

Stromatolites: Laminated (sometimes domed) structures in rocks caused by the trapping and binding of sediment particles by microorganisms.

Succession: A directional change in ecological communities over time, most often used in the context of vegetation change (see Moore, 2001 for an excellent short account; also chapter 9 in Wilkinson 2021a).

Supernova: An 'exploding' star caused by contraction of its core (see chapter 3 in Narlikar, 1999).

Symbiosis: In its original usage this described two organisms living very closely together. Such a relationship could be mutually beneficial, harmful to one of the partners, or neutral with respect to benefits. This is the way the term is used in this book; however, some authors use it as a synonym for mutualism (*q.v.*) (see Wilkinson, 2001b for a discussion of these alternative definitions).

Teleological: Explanations that assume a purpose—for example explanations of an organism's adaptations as designed for the purpose by a God. Such explanations are rightly considered highly suspect by science; suggestions that ideas like Gaia (*q.v.*) and maximum entropy production (section 2.4) imply such a purpose have made many scientists treat them with suspicion.

Testate amoebae: Shell-building protozoa which are often common in soils and freshwater, especially in sediments of high organic matter content (see Figure 4.1 and Wilkinson 2022).

Thermodynamics: The study of energy transformations (see Pielou, 2001 for a simple introduction to the thermodynamics of the Earth system).

Xylem: Vascular woody plant tissue that allows the transport of water—along with dissolved minerals—around a plant.

The geological timescale

The geological timescale for Earth is split into four 'eons', which are subdivided into 'eras'; these eras are further subdivided into the more familiar geological periods (systems). Exact dates vary between different sources; I have updated the version used in the first edition of this book mainly following Boenigk et al. (2015).

Eons—see also within the glossary above:

- *Hadean*: before approximately 4,000 million years ago.
- *Archaean*: from approximately 4,000 to 2,500 million years ago.
- *Proterozoic*: from 2,500 to 541 million years ago.
- *Phanerozoic*: from 541 million years ago until the present.

The Phanerozoic in more detail

Within this book I only use formal names of eras and periods (systems) within the most recent geological era—the Phanerozoic. Note that the well-known period of the 'Tertiary' no longer has formal status and is split into the Palaeogene and the Neogene—this doesn't stop many older geologists still talking about the 'Cretaceous/Tertiary boundary' when discussing mass extinctions, etc., as old habits die hard. The subdivisions of the Phanerozoic are shown in the following table; except for the recent Quaternary, all dates have been rounded to the nearest million years.

Era	Period (System)	Age (millions of years ago)
Cenozoic	Quaternary	0–2.58
Cenozoic	Neogene	2.58–23
Cenozoic	Palaeogene	23–66
Mesozoic	Cretaceous	66–145
Mesozoic	Jurassic	145–201
Mesozoic	Triassic	201–252
Palaeozoic	Permian	252–298
Palaeozoic	Carboniferous	298–359
Palaeozoic	Devonian	359–419
Palaeozoic	Silurian	419–443
Palaeozoic	Ordovician	443–485
Palaeozoic	Cambrian	485–541

References

Abraham, E.R., Law, C.S., Boyd, P.W., Lavender, S.J., Maldonado, M.T. and Bowie, A.R. 2000. Importance of stirring in the development of an iron-fertilized phytoplankton bloom. *Nature* **407**, 727–730.

Ackland, G.J., Clark, M.A. and Lenton, T.M. 2003. Catastrophic desert formation in Daisyworld. *Journal of Theoretical Biology* **223**, 39–44.

Adams, B., Carr, J., Lenton, T.M. and White, A. 2003. One-dimensional Daisyworld: spatial interactions and pattern formation. *Journal of Theoretical Biology* **233**, 505–513.

Adams, F.C. and Napier, K.J. 2022. Transfer of rocks between planetary systems: Panspermia revisited. *Astrobiology* **22**, 1429–1442.

Aleksander, I. 2002. Understanding information, bit by bit. In: *It Must Be Beautiful; Great Equations of Modern Science*. (ed. Farmelo, G.), pp 131–148. London: Granta Books.

Allee, W.C., Emerson, A.E., Park, O. and Schmidt, K.P. 1949. *Principles of Animal Ecology*. Philadelphia: W.B. Saunders.

Allen, A.P., Gillooly, J.F. and Brown, J.H. 2005. Linking the global carbon cycle to individual metabolism. *Functional Ecology* **19**, 202–213.

Anderson, R.M. and May, R.M. 1991. *Infectious Diseases of Humans: Dynamics and Control*. Oxford: Oxford University Press.

Andrews, J.H. 2017. *Comparative Ecology of Microorganisms and Macroorganisms*, 2nd ed. New York: Springer.

Antcliffe, J.B., Liu, A.G., Menon, L.R., McIlroy, D., McLoughlin, N. and Wacey, D. 2017. Understanding ancient life: how Martin Brasier changed the way we think about the fossil record. In: *Earth System Evolution and Early Life: A Celebration of the Work of Martin Brasier* (eds Brasier, A.T., McIlroy, D. and McLoughlin, N.), pp 19–31. Geological Society Special Publication No. 448. London: Geological Society.

Archibold, O.W. 1995. *Ecology of World Vegetation*. London: Chapman and Hall.

Arrigo, K.R. 2005. Marine microorganisms and global nutrient cycling. *Nature* **437**, 349–355.

Arroyo-Mora, J.P., Sánchez-Azofeifa, G.A., Rivard, B., Calvo, J.C. and Janzen, D.H. 2005. Dynamics in landscape structure and composition for the Chorotega region, Costa Rica from 1960 to 2000. *Agriculture, Ecosystems and Environment* **106**, 27–39.

Arthur, R. and Nicholson, A. 2017. An entropic model of Gaia. *Journal of Theoretical Biology* **430**, 177–184.

Arthur, R. and Nicholson, A. 2022. Selection principles for Gaia. *Journal of Theoretical Biology* **533**, 110940.

Ashmole, P. and Ashmole M. 2000. *St Helena and Ascension Island: a natural history*. Oswestry: Anthony Nelson.

Atkinson, R.J.C. 1957. Worms and weathering. *Antiquity* **31**, 219–233.

Atlas, R.M. and Bartha, R. 1998. *Microbial Ecology: Fundamentals and Applications*, 4th ed. Menlo Park, CA: Benjamin Cummins.

Attenborough, D. and Hughes, J. 2020. *A Life on Our Planet*. London: Witness Books.

Ayres, P. 2012. *Shaping Ecology: The Life of Arthur Tansley*. Chichester: Wiley-Blackwell.

Backhaus, T., Meeßen, J., Demets, R., de Vera, J.-P. and Ott, S. 2019. Characterization of viability of the lichen *Buellia frigida* after 1.5 years in space on the international space station. *Astrobiology* **19**, 233–241.

Ball, P. 2005a. Seeking the solution. *Nature* **436**, 1084–1085.

Ball, P. 2005b. *Elegant Solutions*. Cambridge: Royal Society of Chemistry.

Barnes, D.K.A. 2002. Invasions by marine life on plastic debris. *Nature* **416**, 808–809.

Bar-On, Y.M., Phillips, R. and Milo, R. 2018. The biomass distribution on Earth. *Proceedings of the National Academy of Sciences of the United States of America* **115**, 6506–6511.

Bateman, I.J., Anderson, K., Argles, A., et al. 2023. A review of planting principles to identify the right place for the right tree for 'net zero plus' woodlands: applying a place-based natural capital framework for sustainable, efficient and equitable (SEE) decisions. *People and Nature* **5**, 271–301.

Battin, T.J., Lauerwald, R., Bernhardt, E.S., et al. 2023. River ecosystem metabolism and carbon biogeochemistry in a changing world. *Nature* **613**, 449–459.

Baumann, E. 1911. Die vegetation des untersees (Bondensee); eine floristich-kritische und biologische studie. *Archive für Hydrobiology Supplement* **1**, 1–554.

Beerling, D.J. 2005. Leaf evolution: gases, genes and geochemistry. *Annals of Botany* **96**, 345–352.

Beerling, D. 2019a. *Making Eden*. Oxford: Oxford University Press.

Beerling D.J. 2019b. Can plants help us avoid seeding a human-made climate catastrophe? *Plants, People and Planet* **1**, 310–314.

Beerling, D.J. and Berner, R.A. 2005. Feedbacks and the coevolution of plants and atmospheric CO_2. *Proceedings of the National Academy of Sciences of the United States of America* **102**, 1302–1305.

Beerling, D.J., Chaloner, W.G., Huntley, B., Pearson, J.A. and Tooley, M.J. 1993. Stomatal density responds to the glacial cycle of environmental change. *Proceedings of the Royal Society London B* **251**, 133–138.

Beerling, D.J., Lomax, B.H., Upchurch Jr, G.R., et al. 2001a. Evidence for the recovery of terrestrial ecosystems ahead of marine primary production following a biotic crisis at the Cretaceous–Tertiary boundary. *Journal of the Geological Society, London* **158**, 737–740.

Beerling, D.J., Osborne, C.P. and Chaloner, W.G. 2001b. Evolution of leaf-form in land plants linked to atmospheric CO_2 decline in the Late Palaeozoic era. *Nature* **410**, 352–354.

Beerling, D.J., Lomax, B.H., Royer, D.L., Upchurch Jr, G.R. and Kump, L.R. 2002. An atmospheric ρCO_2 reconstruction across the Cretaceous–Tertiary boundary from leaf megafossils. *Proceedings of the National Academy of Sciences of the United States of America* **99**, 7836–7840.

Beerling D.J., Kantzas, E.P., Lomas, M.R., et al. 2020. Potential for large-scale CO_2 removal via enhanced rock weathering with croplands. *Nature* **583**, 242–248.

Begon, M. and Townsend, C.R. 2021. *Ecology: Individuals, Populations and Communities*, 5th ed. Oxford: Wiley-Blackwell.

Bennett, K.D. 1989. A provisional map of forest types for the British Isles 5000 years ago. *Journal of Quaternary Science* **4**, 141–144.

Bergman, N.M., Lenton, T.M. and Watson, A.J. 2004. COPSE: a new model of biogeochemical cycling over Phanerozoic time. *American Journal of Science* 304, 397–437.

Berkner, L.V. and Marshall, L.C. 1972. Oxygen and evolution. In: *Understanding the Earth*, 2nd ed. (eds Glass, I.G., Smith, P.J. and Wilson, R.C.L.), pp 143–149. Cambridge, MA: MIT Press.

Berner, R.A. 1998. The carbon cycle and CO_2 over Phanerozoic time: the role of land plants. *Philosophical Transactions of the Royal Society London B* **353**, 75–82.

Berner, R.A. 2004. *The Phanerozoic Carbon Cycle*. Oxford: Oxford University Press.

Berry, J.A., Beerling, D.J. and Franks, P.J. 2010. Stomata: key players in the Earth system, past and present. *Current Opinion in Plant Biology* **13**, 233–240.

Bessa-Gomes, C., Legendre, S. and Clobert, J. 2004. Allee effect, mating systems and the extinction risk in populations with two sexes. *Ecology Letters* **7**, 802–812.

Betts, R.A. 2000. Offset of the potential carbon sink from boreal forestation by decreases in surface albedo. *Nature* **408**, 187–190.

Betts, R.A. 2004. Global vegetation and climate: self-beneficial effects, climate forcings and climate feedbacks. *Journal de Physique IV France* 121, 37–60.

Betts, R.A. and Lenton, T.M. 2008. Second chances for lucky Gaia: a hypothesis of sequential selection. *Hadley Centre Technical Note* 77.

Biggs, T.E.G., Huisman, J. and Brussaard, C.P.D. 2021. Viral lysis modifies seasonal phytoplankton dynamics and carbon flow in the Southern Ocean. *The ISME Journal* **15**, 3615–3622.

Birks, H.J.B. 1989. Holocene isochrones maps and patterns of tree-spreading in the British Isles. *Journal of Biogeography* **16**, 503–540.

Birks, H.J.B. 2019. Contributions of Quaternary botany to modern ecology and biogeography. *Plant Ecology and Diversity* **12**, 189–385.

Boenigk, J., Wodniok, S. and Glücksman, E. 2015. *Biodiversity and Earth History*. Berlin: Springer.

Bogen, J. 1995. Teleological explanation. In: *The Oxford Companion to Philosophy* (ed. Hondrich, T.), pp 868–869. Oxford: Oxford University Press.

Bond, W.J. 2019. *Open Ecosystems: Ecology and Evolution Beyond the Forest Edge*. Oxford: Oxford University Press.

Bond, W.J. and Keeley, J.E. 2005. Fire as a global 'herbivore': the ecology and evolution of flammable ecosystems. *Trends in Ecology and Evolution* **20**, 387–394.

Bonen, L. and Doolittle, W.F. 1976. Partial sequences of 16S rRNA and the phylogeny of blue-green algae and chloroplasts. *Nature* **261**, 669–673.

Bower, F.O. 1930. *Size and Form in Plants*. London: Macmillan and Co.

Bradshaw, R.H.W. 1999. Spatial response of animals to climate change during the Quaternary. *Ecological Bulletins* **47**, 16–21.

Brasier, M. 2009. *Darwin's Lost World*. Oxford: Oxford University Press.

Brasier, M.D. 1979. The early fossil record. *Chemistry in Britain* **15**, 588–592.

Brasier, M.D., Green, O.R., Jephcoat, A.P., et al. 2002. Questioning the evidence for Earth's oldest fossils. *Nature* **416**, 76–81.

Brook, B.W. and Kikkawa, J. 1998. Examining threats faced by island birds: a population viability analysis on the Capricorn Silvereye using long-term data. *Journal of Applied Ecology* **35**, 491–503.

Brown, R. and Bauer, F. 1814. *General Remarks, Geographical and Systematical, on the Botany of Terra Australis*. Published as Appendix to Flinders, M. *A Voyage to Terra Australis*. London: W. Bulmer.

Buckling, A. and Rainey, P.B. 2002. The role of parasites in sympatric and allopatric host diversification. *Nature* **420**, 496–499.

Bush, M.B. and Colinvaux, P.A. 1994. Tropical forest disturbance: paleoecological records from Darien, Panama. *Ecology* **75**, 1761–1768.

Busher, P.E. 2016. Beavers. In: *Handbook of Mammals of the World*. Vol. 6 (eds Wilson, D.E., Lacher Jr, T.E. and Mittermeier, R.A.), pp 150–168. Barcelona: Lynx Edicions.

Butterfield, H. 1957. *The Origins of Modern Science, 1300–1800*, 2nd ed. London: G. Bell & Sons.

Calow, P. (ed) 1999. *Concise Encyclopaedia of Ecology*. Oxford: Blackwell.

Canfield, D.E. 1999. A breath of fresh air. *Nature* **400**, 503–505.

Canfield, D.E. 2014. *Oxygen*. Princeton: Princeton University Press

Cano, R.J. and Borucki, M.K. 1995. Revival and identification of bacterial spores in 25- to 40-million-year-old Dominican Amber. *Science* **268**, 1060–1064.

Capone, D.G., Burns, J.A., Monotoya, J.P., et al. 2005. Nitrogen fixation by *Trichodesmium* spp: an important source of new nitrogen to the tropical and subtropical Atlantic Ocean. *Global Biogeochemical Cycles* **19**, GB2024.

Carr, B.J. and Rees, M.J. 1979. The anthropic principle and the structure of the physical world. *Nature* **278**, 605–612.

Carter, R.N. and Prince, S.D. 1981. Epidemic models used to explain biogeographical distribution limits. *Nature* **293**, 644–645.

Case, T.J. 2000. *An Illustrated Guide to Theoretical Ecology*. Oxford: Oxford University Press.

Catling, D.C. and Claire, M.W. 2005. How Earth's atmosphere evolved to an oxic state: a status report. *Earth and Planetary Science Letters* **237**, 1–20.

Caughley, G. 1994. Directions in conservation biology. *Journal of Animal Ecology* **63**, 215–244.

Cavalier-Smith, T. 2009. Predation and eukaryote cell origins: a coevolutionary perspective. *The International Journal of Biochemistry and Cell Biology* **41**, 307–322.

Cavalier-Smith T. 2013. Symbiogenesis: mechanisms, evolutionary consequences, and systematic implications. *Annual Review of Ecology, Evolution and Systematics* **44**, 145–172.

Cazzolla Gatti, R. 2018. Is Gaia alive? The future of a symbiotic planet. *Futures* **104**, 91–99.

Cazzolla Gatti, R., Fath, B., Hordijk, W., Kauffman, S. and Ulanowicz, R. 2018. Niche emergence as an autocatalytic process in the evolution of ecosystems. *Journal of Theoretical Biology* **454**, 110–117.

Chambers, F.M. 1995. Climate response, migrational lag, and pollen representation: the problems posed by Rhododendron and Acer. *Historical Biology* **9**, 243–256.

Chambers, F.M. and Brain, S.A. 2002. Paradigm shifts in late-Holocene climatology? *The Holocene* **12**, 239–249.

Chapman, E.J., Childers, D.L. and Vallio, J.J. 2016. How the second law of thermodynamics has informed ecosystem ecology through its history. *Bioscience* 66, 27–39.

Charlson, R.J., Lovelock, J.E., Andreae, M.O. and Warren, S.G. 1987. Oceanic phytoplankton, atmospheric sulphur, cloud albedo and climate. *Nature* **326**, 655–661.

Chen, D., Zhang, Q., Tang, W., et al. 2020. The evolutionary origin and domestication of goldfish (*Carassius auratus*). *Proceedings of the National Academy of Sciences of the United States of America* **117**, 29775–29785.

Cherrett, J.M. 1989. Key concepts: the results of a survey of our members' opinions. In: *Ecological Concepts* (ed. Cherrett, J.M.), pp 1–16. Oxford: Blackwell.

Chopra, A. and Lineweaver, C.H. 2016. The case of a Gaian bottleneck: the biology of habitability. *Astrobiology* **16**, 7–22.

Christenhusz, M.J.M., Fay, M.F. and Chase, M.W. 2018. *Plants of the World*. Richmond: Royal Botanic Gardens Kew.

Chyba, C.F. 2000. Energy for microbial life on Europa. *Nature* **403**, 381–382.

Chyba, C.F. and Phillips, C.B. 2001. Possible ecosystems and the search for life on Europa. *Proceedings of the National Academy of Sciences of the United States of America* **98**, 801–804.

Clark, D.R., Ferguson, R.M.W., Harris, D.N., et al. 2018. Streams of data from drops of water: 21st century molecular microbial ecology. *WIREs Water* **2018**, e1280.

Clarke, A. 2017. *Principles of Thermal Ecology*. Oxford: Oxford University Press.

Cleland, C.E. 2020. Is it possible to scientifically reconstruct the history of life on Earth? In: *Philosophy of Science for Biologists* (eds Kampoutakis, K. and Uller, T.), pp 193–215. Cambridge: Cambridge University Press.

Clymo, R.S., Turunen, J. and Tolonen, K. 1998. Carbon accumulation in peatland. *Oikos* **81**, 368–388.

Cockell, C.S. and Blaustein, A.R. 2000. 'Ultraviolet spring' and the ecological consequences of catastrophic impacts. *Ecology Letters* **3**, 77–81.

Coe, M. 1987. Unforeseen effects of control. *Nature* **327**, 367.

Cohen, H.F. 2015. *The Rise of Modern Science Explained. A Comparative History*. Cambridge: Cambridge University Press.

Cohen, J.E. 1995. *How Many People Can the Earth Support?* New York: W.W. Norton.

Cohen, J.E. and Tilman, D. 1996. Biosphere 2 and biodiversity: the lessons so far. *Science* **274**, 1150–1151.

Cohen, J.E., Pimm, S.L., Yodzis, P. and Saldana, J. 1993. Body sizes of animal predators and animal prey in food webs. *Journal of Animal Ecology* **62**, 67–78.

Cohen, J.E., Jonsson, T. and Carpenter, S.R. 2003. Ecological community description using the food web, species abundance, and body size. *Proceedings of the National Academy of Sciences of the United States of America* **100**, 1781–1786.

Cohen, P.A. and Kodner, R.B. 2022. The earliest history of eukaryotic life: uncovering an evolutionary story through the integration of biological and geological data. *Trends in Ecology and Evolution* **37**, 246–256.

Coles, B. 2001. The impact of beaver activity on stream channels: some implications for past landscapes and human activity. *Journal of Wetland Archaeology* **1**, 55–82.

Colinvaux, P. 1980. *Why Big Fierce Animals Are Rare*. London: Penguin.

Colinvaux, P. 1993. *Ecology 2*. New York: Wiley.

Colinvaux, P. 2007. *Amazon Expeditions*. New Haven: Yale University Press.

Colman, D.R., Poudel, S., Stamps, B.W., Boyd, E.S. and Spear, J.R. 2017. The deep hot biosphere: twenty-five years of retrospection. *Proceedings of the National Academy of Sciences of the United States of America* **114**, 6895–6903.

Condamine, F.L., Guinot, G., Benton, M.J. and Currie, P.J. 2021. Dinosaur biodiversity declined well before the asteroid impact, influenced by ecological and environmental pressure. *Nature Communications* **12**, 3833.

Connell, J.H. 1961. The influence of interspecific competition and other factors on the distribution of the barnacle Chthamalus stellatus. *Ecology* **42**, 710–723.

Conway Morris, S. 1989. Burgess shale faunas and the Cambrian explosion. *Science* **246**, 339–345.

Conway Morris, S. 1998. *The Crucible of Creation*. Oxford: Oxford University Press.

Conway Morris, S. 2000. The Cambrian 'explosion': slow-fuse or megatonnage? *Proceedings of the National Academy of Sciences of the United States of America* **97**, 4426–4429.

Conway Morris, S., Peel, J.S., Higgins, A.K., Soper, N.J. and Davis, N.C. 1987. A Burgess shale-like fauna from the Lower Cambrian of North Greenland. *Nature* **326**, 181–183.

Coombs, F. 1978. *The Crows: A Study of the Corvids of Europe*. London: B.T. Batsford.

Cott, H.B. 1940. *Adaptive Coloration in Animals*. London: Methuen.

Coulson, T., Mace, G.M., Hudson, E. and Possingham, H. 2001. The use and abuse of population viability analysis. *Trends in Ecology and Evolution* **16**, 219–221.

Courchamp, F., Clutton-Brock, T. and Grenfell, B. 1999. Inverse density dependence and the Allee effect. *Trends in Ecology and Evolution* **14**, 405–410.

Cox, C.B., Moore, P.D. and Ladle, R.J. 2016. *Biogeography: An Ecological and Evolutionary Approach*, 9th ed. Chichester: Wiley-Blackwell.

Cox, P.M., Betts, R.A., Jones, C.D., Spall, S.A. and Totterdale, L.J. 2000. Acceleration of global warming due to carbon-cycle feedbacks in a coupled climate model. *Nature* **408**, 184–187.

Cramp, S. and Perrins, C.M. (eds) 1994. *The Birds of the Western Palearctic, Vol 8*. Oxford: Oxford University Press.

Cramp, S. and Simmons, K.E.L. (eds) 1979. *The Birds of the Western Palaearctic, Vol. 2*. Oxford: Oxford University Press.

Crane, P.R. 2019. An evolutionary and cultural biography of ginkgo. *Plants, People, Planet* **1**, 32–37.

Crawford, R.M.M. 1989. *Studies in Plant Survival*. Oxford: Blackwell Science.

Crick, F. 1982. *Life Itself*. London: Macdonald & Co.

Crist, E. 2004. Concerned with trifles? A geophysiological reading of Charles Darwin's last book. In: *Scientists Debate Gaia* (eds Schneider, S.H., Miller, J.R., Crist, E. and Boston, P.J.), pp 161–172. Cambridge, MA: MIT Press.

Cromsigt, J.P.G.M. and Olff, H. 2008. Dynamics of grazing lawn formation: an experimental test of the role of scale-dependent processes. *Oikos* **117**, 1444–1452.

Cromsigt, J.P.G.M. and te Beest, M. 2014. Restoration of a megaherbivore: landscape-level impacts of white rhinoceros in Kruger National Park, South Africa. *Journal of Ecology* **102**, 566–575.

Cross, J.R. 1975. *Rhododendron ponticum* (biological flora of the British Isles). *Journal of Ecology* **63**, 345–364.

Crutzen, P.J. 2002. Geology of mankind. *Nature* **415**, 23.

Cullen, J. 2011. Naturalised rhododendrons widespread in Great Britain and Ireland. *Hanburyana* **5**, 11–29.

Cunliffe, B 2015. *By Steppe, Desert, and Ocean: The Birth of Eurasia*. Oxford: Oxford University Press.

Currie, W.S. 2011. Units of nature or processes across scales? The ecosystem concept at age 75. *New Phytologist* **190**, 21–34.

Cushman, J.C. 2004. C_3 photosynthesis to crassulacean acid metabolism shift in *Mesembryanthenum crystallinum*: a stress tolerance mechanism. In: *Encyclopaedia of Plant and Crop Science* (ed. Goodman, R.M.), pp 241–244. New York: Marcel Dekker.

Dang, C.K., Chauvert, E. and Gessner, M.O. 2005. Magnitude and variability of process rates in fungal diversity–litter decomposition relationships. *Ecology Letters* **8**, 1129–1137.

Darwin, C. 1859. *On the Origin of Species by Means of Natural Selection*. London: John Murray.

Darwin, C. 1875. *Insectivorous Plants*. London: John Murray.

Darwin, C. 1881. *The Formation of Vegetable Mould Through the Action of Worms, with Observations on Their Habits*. London: John Murray.

Davies, K.M. 2004. Phenylpropanoids. In: *Encyclopaedia of Plant and Crop Science* (ed. Goodman, R.M.), pp 868–871. New York: Marcel Dekker.

Davies, P. 1995. *Are We Alone?* London: Penguin.

Davies, P. 1998. *The Fifth Miracle*. London: Allen Lane.

Davies, P. 2005. A quantum recipe for life. *Nature* **437**, 819.

Davis, S.R. and Wilkinson, D.M. 2004. The conservation management value of testate amoebae as 'restoration' indicators: speculations based on two damaged raised mires in northwest England. *The Holocene* **14**, 135–143.

Dawkins, R. 1982. *The Extended Phenotype*. Oxford: Oxford University Press.

De Castro, F. and Bolker, B. 2005. Mechanisms of disease-induced extinction. *Ecology Letters* **8**, 117–126.

Decloitre, L. 1953. *Researches sur les Rhizopods Thécamoebiens d'A.O.F.* Mémoires de l'Institut Français d'Afrique Noire No. 31. Dakar: IFAN.

Dehnen-Schmutz, K., Perring, C. and Williamson, M. 2004. Controlling *Rhododendron ponticum* in the British Isles: an economic analysis. *Journal of Environmental Management* **70**, 323–332.

de la Bédoyère, G. 2002. *Gods with Thunderbolts: Religion in Roman Britain*. Stroud: Tempus.

del Hoyo, J., Elliot, A. and Sargatal, J. (eds) 1992. *Handbook of the Birds of the World. Vol 1*. Barcelona: Lynx Editions.

del Hoyo, J., Elliot, A. and Sargatal, J. (eds) 1994. *Handbook of the Birds of the World. Vol 2*. Barcelona: Lynx Editions.

De Souza Mendonça Jr, M. 2014. Spatial ecology goes to space: metabiospheres. *Icarus* **233**, 348–351.

Dewar, R.C. 2010. Maximum entropy production as an inference algorithm that translates physical assumptions into macroscopic predictions: don't shoot the messenger. *Entropy* **11**, 931–944

Diamond, J. 1997. *Guns, Germs and Steel*. London: Jonathan Cape.

Diamond, J. 2002. Evolution, consequences and future of plant and animal domestication. *Nature* **418**, 700–707.

Doak, D.F., Bigger, D., Harding, E.K., Marvier, M.A., O'Mally, R.E. and Thomson, D. 1998. The statistical inevitability of stability–diversity relationships in community ecology. *American Naturalist* **151**, 264–276.

Dobbs, B.J.T. 1982. Newton's alchemy and his theory of matter. *Isis* **73**, 511–528.

Dobson, C.M. 2004. Chemical space and biology. *Nature* **432**, 824–828.

Donaldson, J.E., Archibald, S., Govender, N., Pollard, D., Luhdo, Z. and Parr, C.L. 2017. Ecological engineering through fire–herbivory feedbacks drives the formation of savanna grazing lawns. *Journal of Applied Ecology* **55**, 225–235.

Doolittle, W.F. 1981. Is nature really motherly? *The CoEvolution Quarterly* **29**, 58–63.

Doolittle, W.F. 2014. Natural selection through survival alone, and the possibility of Gaia. *Biology and Philosophy* **29** 415–423.

Doolittle, W.F. 2017. Making the most of clade selection. *Philosophy of Science* **84**, 275–295.

Doolittle, W.F. 2019. Making evolutionary sense of Gaia. *Trends in Ecology and Evolution* **34**, 889–894.

Downing, K.L. 2002. The simulated emergence of distributed environmental control in evolving microcosms. *Artificial Life* **8**, 123–153.

Downing, K.L. 2004. Gaia in the machine: the artificial life approach. In: *Scientists Debate Gaia* (eds Schneider, S.H., Miller, J.R., Crist, E. and Boston, P.J.), pp 267–280. Cambridge, MA: MIT Press.

Downing, K. and Zvirinsky, P. 1999. The simulated evolution of biochemical guilds: reconciling Gaia theory and natural selection. *Artificial Life* **5**, 291–318.

Du Rietz, G.E. 1957. Linne som myrforskare. *Uppsala Universitets Arsskrift* **5**, 1–80.

Dusenbery, D.B. 2009. *Living at Micro Scale*. Cambridge, MA: Harvard University Press.

Dutreuil, S. 2018. James Lovelock's Gaia hypothesis. In: *Dreamers, Visionaries, and Revolutionaries in the Life Sciences* (eds Harman, O. and Dietrich, M.R.), pp 272–287. Chicago: University of Chicago Press.

Dyer, J.M. 1995. Assessment of climatic warming using a model of forest species migration. *Ecological Modelling* **79**, 199–219.

Dyke, J. 2021. *Fire, Storm and Flood: The Violence of Climate Change*. London: Head of Zeus.

Dyke, J.G. and Weaver, I.S. 2013. The emergence of environmental homeostasis in complex ecosystems. *PLoS Computational Biology* **9**, e1003050.

Edlin, H.L. 1966. *Trees, Woods and Man*, 2nd ed. London: Collins.

Edwards, D., Cherns, L. and Raven, J.A. 2015. Could land-based early photosynthesizing ecosystems have bioengineered the plant in mid-Palaeozoic times? *Palaeontology* **58**, 803–837.

Ehrlich, P. 1991. Coevolution and its applicability to the Gaia hypothesis. In: *Scientists on Gaia* (eds Schneider, S.H. and Boston, P.J.), pp 19–22. Cambridge, MA: MIT Press.

Eigen, M. and Schuster, P. 1977. The hypercycle. A principle of natural self-organisation, part A: emergence of the hypercycle. *Naturwissenschaften* **58**, 465–523.

Elton, C. 1927. *Animal Ecology*. London: Sidgwick and Jackson.

Elton, C.S. 1958. *The Ecology of Invasion by Animals and Plants*. London: Chapman and Hall.

Eme, L. and Doolittle, W.F. 2015. Primers: archaea. *Current Biology* **25**, R845–R875.

Erwin, T.L. 1982. Tropical forests: their richness in Coleoptera and other arthropod species. *Coleopterists Bulletin* **36**, 74–75.

Falkowski, P.G. 2002. The ocean's invisible forest. *Scientific American* **287**, 38–45.

Falkowski, P.G. and Davis, C.S. 2004. Natural proportions. *Nature* **431**, 131.

Fenchel, T. 1992. What can ecologists learn from microbes: life beneath a square centimetre of sediment surface. *Functional Ecology* **6**, 499–507.

Fenchel, T. 2005. Cosmopolitan microbes and their 'cryptic' species. *Aquatic Microbial Ecology* **41**, 49–54.

Fenchel, T. and Finlay, B.J. 1995. *Ecology and Evolution in Anoxic Worlds*. Oxford: Oxford University Press.

Fenchel, T., King, G.M. and Blackburn, T.H. 1998. *Bacterial Biogeochemistry*, 2nd ed. San Diego: Academic Press.

Finlay, B.J. 2002. Global dispersal of free-living microbial eukaryote species. *Science* **296**, 1061–1063.

Finlay, B.J. and Clarke, K.J. 1999. Ubiquitous dispersal of microbial species. *Nature* **400**, 828.

Finlay, B.J., Marberly, S.C. and Cooper, J.I. 1997. Microbial diversity and ecosystem function. *Oikos* **80**, 209–213.

Finlay, B.J., Esteban, G.F. and Fenchel, T. 2004. Protist diversity is different? *Protist* **155**, 15–22.

Fisher, R.A. 1930. *The Genetical Theory of Natural Selection*. Oxford: The Clarendon Press.

Floate, K.D., Wardhaugh, K.G., Boxall, A.B.A. and Sherratt, T.N. 2005.Fecal residues of veterinary parasiticides: nontarget effects in the pasture environment. *Annual Review of Entomology* **50**, 153–179.

Foissner, W. 1998. An updated compilation of world soil ciliates (Protozoa, Ciliophora), with ecological notes, new records, and descriptions of new species. *European Journal of Protistology* **34**, 195–235.

Foissner, W. 1999. Protist diversity: estimates of the near-imponderable. *Protist* **150**, 363–368.

Fontaneto, D. (ed.) 2011. *Biogeography of Microorganisms: Is Everything Everywhere?* Cambridge: Cambridge University Press.

Förster, B. and Pogson, B.J. 2004. Carotenoids in photosynthesis. In: *Encyclopaedia of Plant and Crop Science* (ed. Goodman, R.M.), pp 245–249. New York: Marcel Dekker.

Foster, D.R. 1992. Land-use history (1730–1990) and vegetation dynamics in central New England, USA. *Journal of Ecology* **80**, 753–772.

Franck, S., Von Bloh, W., Bounana, C. and Schellnhuber, H.-J. 2004. Extraterrestrial Gaias. In: *Scientists Debate Gaia* (eds Schneider, S.H., Miller, J.R., Crist, E. and Boston, P.J.), pp 309–319. Cambridge, MA: MIT Press.

Frank, S.A. 1998. *Foundations of Social Evolution*. Princeton: Princeton University Press.

Fuhrman, J. 2003. Genome sequences from the sea. *Nature* **424**, 1001–1002.

Fuller, E. 2000. *Extinct Birds*. Oxford: Oxford University Press.

Gallego Sala, A.V., Charman, D.J., Harrison, S.P., Li, G. and Prentice, I.C. 2016. Climate-driven expansion of blanket bogs in Britain during the Holocene. *Climates of the Past* **12**, 129–136.

Gardner, D.K. 2018. *Environmental Pollution in China: What Everyone Needs to Know*. Oxford: Oxford University Press.

Gattuso, J.-P. and Buddemeier, R.W. 2000. Calcification and CO_2. *Nature* **407**, 311–313.

Gilbert, D., Amblard, C., Bourdier, G., Francez, A.-J. and Mitchell, E.A.D. 2000. Le régime alimentaire des Thécamoebiens (Protista, Sarcodina). *L'Année Biologique* **39**, 57–68.

Gilbert, P.M. and Mitra, A. 2022. From webs, loops, shunts, and pumps to microbial multitasking: Evolving concepts of marine microbial ecology, the mixoplankton paradigm, and implications for a future ocean. *Limnology and Oceanography* **67**, 585–597.

Givnish, T.J. 1987. Comparative studies of leaf form: assessing the relative roles of selective pressures and phylogenetic constraints. *New Phytologist* **106** (suppl), 131–160.

Gobat, R., Hong, S.E., Snaith, O. and Hong, S. 2021. Panspermia in a Milky Way-like galaxy. *The Astrophysical Journal* **921** (157), 1–16.

Godwin, H. 1975. *History of the British Flora*, 2nd ed. Cambridge: Cambridge University Press.

Gold, T. 1999. *The Deep Hot Biosphere*. New York: Springer.

Goodwin, D. 1986. *Crows of the World*, 2nd ed. London: British Museum (Natural History).

Gould, S.J. 1990. *Wonderful Life*. London: Hutchinson Radius.

Gould, S.J. 1995. 'What is life?' as a problem in history. In: *What Is Life? The Next Fifty Years* (eds Murphy, M.P. and O'Neill, L.A.J.), pp 25–39. Cambridge: Cambridge University Press.

Gould, S.J. 2002. *The Structure of Evolutionary Theory*. Cambridge, MA: Harvard University Press.

Grace, J. 1997. Plant water relations. In: *Plant Ecology*, 2nd ed. (ed. Crawley, M.J.), pp 28–50. Oxford: Blackwell.

Grace, J. 2004. Understanding and managing the global carbon cycle. *Journal of Ecology* **92**, 189–202.

Grace, J. 2019. Has ecology grown up? *Plant Ecology and Diversity* **12**, 387–405.

Grady, M.M., Hutchison, R., McCall, G.J.H. and Rothery, D.A. 1998. *Meteorites: Flux with Time and Impact effects*. Geological Society Special Publication 140. London: Geological Society.

Green, A.J., Baltzinger, C. and Lovas-Kiss, Á. 2022. Plant dispersal syndromes are unreliable, especially for predicting zoochory and long-distance dispersal. *Oikos* **2022**, e08327.

Gregory, J. 2005. *Fred Hoyle's Universe*. Oxford: Oxford University Press.

Gribaldo, S. and Philippe, H. 2002. Ancient phylogenetic relationships. *Theoretical Population Biology* **61**, 391–408.

Gribbin, J. 2000. *Stardust*. London: Allen Lane.

Grime, J.P. 1974. Vegetation classification by reference to strategies. *Nature* **250**, 26–31.

Grime, J.P. 1997. Biodiversity and ecosystem function: the debate deepens. *Science* **277**, 1260–1261.

Grime, J.P. 1998. Benefits of plant diversity to ecosystems: immediate, filter and founder effects. *Journal of Ecology* **86**, 902–910.

Grime, J.P. 2001. *Plant Strategies, Vegetation Processes and Ecosystem Properties*. Chichester: John Wiley.

Grodwohl, J.-B., Porto, F. and El-Hani, C.N. 2018. The instability of field experiments: building an experimental research tradition on the rock shore (1950–1985). *History and Philosophy of the Life Sciences* **40**, 45.

Grubb, P.J. 2016. Trade-offs in interspecific comparisons in plant ecology and how plants overcome proposed constraint. *Plant Ecology and Diversity* **9**, 3–33.

Haila, Y. 1999a. Socioecologies. *Ecography* **22**, 337–348.

Haila, Y. 1999b. Biodiversity and the divide between culture and nature. *Biodiversity and Conservation* **8**, 165–181.

Haldane, J.B.S. 1932. *The Causes of Evolution*. London: Longmans, Green & Co.

Hale, W.G., Margham, J.P. and Saunders, V.A. 2005. *Collins Dictionary of Biology*, 3rd ed. London: Collins.

Hall, S.J. and Raffaelli, D. 1991. Food-web patterns: lessons from a species rich web. *Journal of Animal Ecology* **60**, 823–841.

Hambler, C. and Canney, S.M. 2013. *Conservation*, 2nd ed. Cambridge: Cambridge University Press.

Hamilton, W.D. 1971. Geometry for the selfish herd. *Journal of Theoretical Biology* **31**, 295–311.

Hamilton, W.D. 1995. Ecology in the large: Gaia and Genghis Khan. *Journal of Applied Ecology* **32**, 415–453.

Hamilton, W.D. 1996. *Narrow Roads of Gene Land*. Vol. 1. Oxford: W.H. Freeman.

Hamilton, W.D. 2001. *Narrow Roads of Gene Land*. Vol. 2. Oxford: Oxford University Press.

Hamilton, W.D. and May, R.M. 1977. Dispersal in stable habitats. *Nature* **269**, 578–581.

Hanski, I. 1998. Metapopulation dynamics. *Nature* **396**, 41–49.

Hanski, I. 2016. *Messages from Islands*. Chicago: Chicago University Press.

Harcourt, A.H. 1995. Population viability estimates: theory and practice for a wild gorilla population. *Conservation Biology* **9**, 134–142.

Harding, R.J. and Pomeroy, J.W. 1996. The energy balance of the winter boreal landscape. *Journal of Climatology* **9**, 2778–2787.

Harding, S.P. and Lovelock, J.E. 1996. Exploiter-mediated coexistence and frequency-dependent selection in a numerical model of biodiversity. *Journal of Theoretical Biology* **182**, 109–116.

Hardy, A. 1956. *The Open Sea: The World of Plankton*. London: Collins.

Hazen, R.M. and Roedder, E. 2001. How old are bacteria from the Permian age? *Nature* **411**, 155.

Heger, T.J., Booth, R.K., Sullivan, M.E., et al. 2011. Rediscovery of *Nebela ansata* (Amoebozoa: Arcellinida) in eastern North America: biogeographical implications. *Journal of Biogeography* **38**, 1897–1906.

Heimann, M. 2005. Charles David Keeling 1928–2005. *Nature* **437**, 331.

Henry, J. 2002. *The Scientific Revolution and the Origins of Modern Science*, 2nd ed. London: Palgrave.

Herringshaw, L.G., Callow, R.H.T. and McIlroy, D. 2017. Engineering the Cambrian explosion: the earliest bioturbators as ecosystem engineers. In: *Earth System Evolution and Early Life: A Celebration of the Work of Martin Brasier* (eds Brasier, A.T., McIlroy, D. and and McLoughlin, N.), pp 369–382. Geological Society Special Publication No. 448. London: Geological Society.

Hibberd, J.M. and Furbank, R.T. 2016. Fifty years of C_4 photosynthesis. *Nature* **538**, 177–178.

Hillebrand, H., Watermann, F., Karez, R. and Berninger, U.-G. 2001. Differences in species richness patterns between unicellular and multicellular organisms. *Oecologia* **126**, 114–124.

Hobbs, R.J., Arico, S., Aronson, J., et al. 2006. Novel ecosystems: theoretical and management aspects of the new ecological world order. *Global Ecology and Biogeography* **15**, 1–7.

Hobbs, R.J., Higgs, E.S. and Hall, C.M. 2013. *Novel Ecosystems*. Chichester, Wiley-Blackwell.

Hodkinson, I.D. 1975. Dry weight loss and chemical changes in vascular plant litter of terrestrial origin, occurring in a beaver pond ecosystem. *Journal of Ecology* **63**, 131–142.

Hodkinson, I.D. and Casson, D. 1991. A lesser predilection for bugs: Hemiptera (Insecta) diversity in tropical rain forests. *Biological Journal of the Linnean Society* **43**, 101–109.

Hodkinson, I.D. and Coulson, S.J. 2004. Are High Arctic terrestrial food chains really that simple? – The Bear Island food web revisited. *Oikos* **106**, 427–431.

Hodkinson, I.D., Coulson, S.J. and Webb, N.R. 2004. Invertebrate community assembly along proglacial chronosequences in the High Arctic. *Journal of Animal Ecology* **73**, 556–568.

Hoffman, L.R., D'Argenio, D.A., MacCross, M.J., Zhang, Z, Jones, R.A. and Miller, S.I. 2005. Aminoglycoside antibiotics induce bacterial biofilm formation. *Nature* **436**, 1171–1175.

Holland, J.H. 1992. Genetic algorithms. *Scientific American* **266** (July), 44–50.

Holmes, A. 1944. *Principles of Physical Geology*. London: Thomas Nelson.

Holmes, E.C. 2011. Plague's progress. *Nature* **478**, 465–466.

Holt, R.D. 2020. Ecology 'through the looking glass': what might be the ecological consequences of stopping mutation? In: *Unsolved Problems in Ecology* (eds Dobson, A., Holt, R.D. and Tilman, D.), pp 92–106. Princeton: Princeton University Press.

Hoyle, F. 1950. *The Nature of the Universe*. Oxford: Basil Blackwell.

Hoyle, F. 1957. *The Black Cloud*. London: William Heinemann.

Hoyle, F. 1964. *Of Men and Galaxies*. Washington: Washington University Press.

Hoyle, F. and Wickramasinghe, N.C. 1999. The universe and life: deductions from the weak anthropic principle. *Astrophysics and Space Science* **268**, 89–102.

Huggett, R.J. 1999. Ecosystem, biosphere, or Gaia? What to call the global ecosystem. *Global Ecology and Biogeography* **8**, 425–431.

Huntley, B. 1994. Plant species' response to climate change: implications for the conservation of European birds. *Ibis* **137**, s127–s138.

Huston, M.A. 1997. Hidden treatments in ecological experiments: re-evaluating the ecosystem function of biodiversity. *Oecologia* **110**, 449–460.

Hutchinson, G.E. 1964. The lacustrine microcosm reconsidered. *American Scientist* **52**, 334–341.

Hutchinson, G.E. 1965. *The Ecological Theatre and the Evolutionary Play*. New Haven: Yale University Press.

Hutchinson G.E. 1970. The biosphere. *Scientific American* **233**, 44–53.

Huxley, T.H. 1887. *Physiography: An Introduction to the Study of Nature*, 3rd ed. London: MacMillan.

Inkpen, S.A. and Doolittle, W.F. 2022. *Can Microbial Communities Regenerate?* Chicago: Chicago University Press.

Jaakkola, S.T., Zerulla, K., Guo, Q., et al. 2014. Halophilic Archean cultivated from surface sterilized Middle–Late Eocene rock salt are polypode. *PLoS ONE* **9**, e110533.

Jackson, R. and Gabric, A. 2022. Climate change impacts on the marine cycling of biogenic sulfur: a review. *Microorganisms* **10**, 1581.

Janzen, D.H. 1986. Lost plants. *Oikos* **46**, 129–131.

Jassey, V.E.J., Walcker, R., Kardol, P., et al. 2022. Contributions of soil algae to the global carbon cycle. *New Phytologist* **234**, 64–76.

Javaux, E.J. 2019. Challenges in evidencing the earliest traces of life. *Nature* **572**, 451–460.

Jepson, P. and Canney, S. 2003. Values-led conservation. *Global Ecology and Biogeography* **12**, 271–274.

Johnson, N.P. and Mueller, J. 2002. Updating the accounts: global mortality of the 1918–1920 'Spanish' influenza pandemic. *Bulletin of the History of Medicine* **76**, 105–115.

Jones, C.G., Lawton, J.H. and Shachak, M. 1994. Organisms as ecosystem engineers. *Oikos* **69**, 373–386.

Jones, C.G., Lawton, J.H. and Shachak, M. 1997. Positive and negative effects of organisms as physical ecosystem engineers. *Ecology* **78**, 1946–1957.

Jones, D. 2005. Personal effects. *Nature* **438**, 14–16.

Kassen, R., Llewellyn, M. and Rainey, P.B. 2004. Ecological constraints on diversification in a model adaptive radiation. *Nature* **431**, 984–988.

Kasting, J. 2010. *How to Find a Habitable Planet*. Princeton: Princeton University Press.

Kasting, J.F. and Siefert, J.L. 2002. Life and the evolution of Earth's atmosphere. *Science* **296**, 1066–1068.

Kasting, J.F., Whitmire, D.P. and Reynolds, R.T. 1993. Habitable zones around main sequence stars. *Icarus* **101**, 108–128.

Kaul, R.B., Kramer, A.M., Dobbs, F.C. and Drake, J.M. 2016. Experimental demonstration of an Allee effect in microbial populations. *Biology Letters* **12**, 20160070.

Kawecki, T.J., Lenski, R.E., Ebert, D., Hollis, B., Olivieri, I. and Whitlock, M.C. 2012. Experimental evolution. *Trends in Ecology and Evolution* **27**, 547–560.

Keeling, P.J. and Koonin E.V. (eds) 2014. *The Origin and Evolution of Eukaryotes*. New York: Cold Spring Harbor Laboratory Press.

Kemp, L., Xu, C., Depledge, J., et al. 2022. Climate endgame: exploring catastrophic climate change scenarios. *Proceedings of the National Academy of Sciences of the United States of America* **119**, e2108146119.

Kennedy, M.J., Reader, S.L. and Swierczynski, L.M. 1994. Preservation records of micro-organisms: evidence of the tenacity of life. *Microbiology* **140**, 2513–1529.

Kenrick, P. and Crane, P.R. 1997. The origin and early evolution of plants on land. *Nature* **389**, 33–39.

Kingsland, S.E. 1995. *Modelling Nature: Episodes in the History of Population Ecology*, 2nd ed. Chicago: Chicago University Press.

Kirchner, J.W. 2003. The Gaia hypothesis: conjectures and refutations. *Climatic Change* **58**, 21–45.

Kirkwood, T.B.L. and Austad, S.N. 2000. Why do we age? *Nature* **408**, 233–238.

Kissel, J. and Krueger, F.R. 1987. The organic component in dust from comet Halley as measured by the PUMA mass spectrometer on board Vega 1. *Nature* **326**, 755–760.

Kleidon, A. 2004. Beyond Gaia: thermodynamics of life and Earth system functioning. *Climatic Change* **66**, 271–319.

Kleidon, A. 2016. *Thermodynamic Foundations of the Earth System*. Cambridge: Cambridge University Press.

Kleidon, A., Fraedrich, K. and Heimenn, M. 2000. A green planet versus a desert world: estimating the maximum effect of vegetation on land surface climate. *Climatic Change* **44**, 471–493.

Klinger, L.F., Taylor, J.A. and Franzen, L.G. 1996. The potential role of peatland dynamics in ice-age initiation. *Quaternary Research* **45**, 89–92.

Krebs, C.J. 2009. *Ecology*, 6th ed. London: Benjamin Cummings.

Kühlbrandt, W. 2001. Chlorophylls galore. *Nature* **411**, 896–899.

Kuhlman, K.R., Allenbach, L.B., Ball, C.L., et al. 2005. Enumeration, isolation, and characterization of ultraviolet (UV-C) resistant bacteria from rock varnish in the Whipple Mountains, California. *Icarus* **174**, 585–595.

Kump, L.R., Kasting, J.F. and Crane, R.C. 2010. *The Earth System*, 3rd ed. San Francisco: Prentice Hall.

Lacy, R.C. 2000. Structure of the VORTEX simulation model for population viability analysis. *Ecological Bulletins* **48**, 191–203.

Lal, R. 2004. Soil carbon sequestration to mitigate climate change. *Geoderma* **123**, 1–22.

Laland, K.N., Odling-Smee, F.J. and Feldman, M.W. 1999. Evolutionary consequences of niche construction and their implications for ecology. *Proceedings of the National Academy of Sciences of the United States of America* **96**, 10242–10247.

Lamentowicz, M., Bragazza, L., Buttler, A., Jassey, V.E.J. and Mitchell, E.A.D. 2013. Seasonal patterns of testate amoebae diversity, community structure and species–environment relationships in four *Sphagnum*-dominated peatlands along a 1300 m altitudinal gradient in Switzerland. *Soil Biology and Biochemistry* **67**, 1–11.

Lan, X., Nisbet, E.G., Dlugokencky, E.J. and Michel, S.E. 2021. What do we know about the global methane budget? Results for four decades of atmospheric CH_4 observation and the way forward. *Philosophical Transactions of the Royal Society A* **379**: 20200440.

Lande, R. 1993. Risks of population extinction from demographic and environmental stochasticity and random catastrophes. *American Naturalist* **142**, 911–927.

Lange, O.L., Kidron, G.J., Budel, B., Meyer, A., Kilian, E. and Abeliovich, A. 1992.Taxonomic composition and photosynthetic characteristics of the 'biological soil-crusts' covering sand dunes in the western Negev Desert. *Functional Ecology* **6**, 519–527.

Lara, E., Heger, T.J., Scheihing, R. and Mitchell, E.A.D. 2011. COI gene and ecological data suggests size-dependent high dispersal and low intra-specific diversity in free-living terrestrial protists (Euglyphida: *Assulina*). *Journal of Biogeography* **38**, 640–650.

Latour, B. 2018. Down to Earth. Cambridge: Polity Press.

Lawton, J. 2001. Earth system science. *Science* **292**, 1965.

Lawton, J.H. 1996. Patterns in ecology. *Oikos* **75**, 145–147.

Lawton, J.H. 1999. Are there general laws in ecology? *Oikos* **84**, 177–192.

Lázaro, A., Gómez-Martínez, C., González-Estévez, M.A. and Hidalgo, M. 2022. Portfolio effect and asynchrony as drivers of stability in plant-pollinator communities along a gradient of landscape heterogeneity. *Ecography* **2022**, e06112.

Leakey, R. and Lewin, R. 1995. *The Sixth Extinction: Biodiversity and its Survival*. London: Weidenfeld and Nicholson.

Leather, S. 2022. *Insects: A Very Short Introduction*. Oxford: Oxford University Press.

Lee, D.W., O'Keefe, J., Holbrook, N.M. and Feild, T.S. 2003. Pigment dynamics and autumn leaf senescence in a New England deciduous forest, eastern USA. *Ecological Research* **18**, 677–694.

Leidy, J. 1879. *Fresh-water Rhizopods of North America*. Washington: Government Printing Office.

Lekevičius, E. 2006. The Russian paradigm in ecology and evolutionary biology: pro et contra. *Acta Zoologica Lituanica* **16**, 3–19.

Lenton, T. 2016. *Earth Systems Science: A Very Short Introduction*. Oxford: Oxford University Press.

Lenton, T. and Watson, A. 2011. *Revolutions that Made the Earth*. Oxford: Oxford University Press.

Lenton, T.M. 1998. Gaia and natural selection. *Nature* **394**, 439–447.

Lenton, T.M. 2001. The role of land plants, phosphorus weathering and fire in the rise and regulation of atmospheric oxygen. *Global Change Biology* **7**, 613–629.

Lenton, T.M. 2004. Clarifying Gaia: regulation with or without natural selection. In: *Scientists Debate Gaia* (eds Schneider, S.H., Miller, J.R., Crist, E. and Boston, P.J.), pp 15–25. Cambridge, MA: MIT Press.

Lenton, T.M. and Daines, S.J. 2017. Matworld—the biogeochemical effects of early life on land. *New Phytologist* **215**, 531–537.

Lenton, T.M. and Huntingford, C. 2003. Global terrestrial carbon storage and uncertainties in its temperature sensitivity examined with a simple model. *Global Change Biology* **9**, 1333–1352.

Lenton, T.M. and Latour, B. 2018. Gaia 2.0. *Science* **361**, 1066–1068.

Lenton, T.M. and Lovelock, J.E. 2001. Daisyworld revisited: quantifying biological effects on planetary self-regulation. *Tellus* **53B**, 288–305.

Lenton, T.M. and Watson, A.J. 2000a. Redfield revisited. 1. Regulation of nitrate, phosphate, and oxygen in the ocean. *Global Biogeochemical Cycles* **14**, 225–248.

Lenton, T.M. and Watson, A.J. 2000b. Redfield revisited 2. What regulates the oxygen content of the atmosphere? *Global Biogeochemical Cycles* **14**, 249–268.

Lenton T.M. and Watson, A.J. 2004. Biotic enhancement of weathering, atmospheric oxygen and carbon dioxide in the Neoproterozoic. *Geophysical Research Letters* **31**, L05202.

Lenton, T.M. and Wilkinson, D.M. 2003. Developing the Gaia theory, a response to the criticisms of Kirchner and Volk. *Climatic Change* **58**, 1–12.

Lenton, T.M., Schellnhuber, H.J. and Szathmáry, E. 2004. Co-evolution: Earth history involves tightly entwined transitions of information and the environment, but where is this process heading? *Nature* **431**, 913.

Lenton, T.M., Daines, S.J., Dyke, J., Nicholson, A.E., Wilkinson, D.M. and Williams, H.T.P. 2018. Selection for Gaia across multiple scales. *Trends in Ecology and Evolution* **33**, 633–645.

Lenton, T.M., Dutreuil, S. and Latour, B. 2020. Life on Earth is hard to spot. *The Anthropocene Review* **7**, 248–272.

Lenton, T.M., Kohler, T.A., Marquet, P.A., et al. 2021. Survival of the systems. *Trends in Ecology and Evolution* **36**, 333–344.

Lepland, A., van Zuilen, M.A., Arrhenius, G., Whitehouse, M.J. and Fedo, C.M. 2005. Questioning the evidence for Earth's earliest life—Akilia revisited. *Geology* **33**, 77–79.

Lever, C. 1977. *The Naturalized Animals of the British Isles*. London: Hutchinson.

Levins, R. 1969. Some demographic and genetic consequences of environmental heterogeneity for biological control. *Bulletin of the Entomological Society of America* **15**, 237–240.

Lev-Yadum, S., Gopher, A. and Abbo, S. 2000. The cradle of agriculture. *Science* **288**, 1602–1603.

Lewontin, R. 1983. The organism as the subject and object of evolution. *Scientia* **118**, 63–82.

Lewontin, R. 2002. *The Triple Helix*. Cambridge, MA: Harvard University Press.

Lindbladh, M. and Bradshaw, R. 1995. The development and demise of a medieval forest-meadow system at Linnaeus' birthplace in southern Sweden: implications for conservation and forest history. *Vegetation History and Archaeobotany* **4**, 153–160.

Lindsey, P.A., Alexander, R., Mills, M.G.L., Romañach, S. and Woodroffe, R. 2007. Wildlife viewing preferences of visitors to protected areas in South Africa: implications for the role of ecotourism in conservation. *Journal of Ecotourism* **6**, 19–33.

Lipman, C.B. 1928. The discovery of living microorganisms in ancient rocks. *Science* **68**, 272–231.

Lisner, A., Konečná, M., Blažek, P. and Lepš, J. 2023. Community biomass is driven by dominants and their characteristics—the insight from a field biodiversity experiment with realistic species loss scenario. *Journal of Ecology* **111**, 240–250.

Liu, L., Yang, J., Yu, Z. and Wilkinson, D.M. 2015. The biogeography of abundant and rare bacterioplankton in the lakes and reservoirs of China. *The ISME Journal* **9**, 2068–2077.

Liu, L., Chen, H., Liu, M., et al. 2019. Response of the eukaryotic plankton community to the cyanobacterial biomass cycle over 6 years in two subtropical reservoirs. *The ISME Journal* **13**, 2196–2208.

Loisel, J., Gallego-Sala, A., Amesbury, M.J., et al. 2021. Expert assessment of future vulnerability of the global peatland carbon sink. *Nature Climate Change* **11**, 70–77.

Lomax, B., Beerling, D., Upchurch Jr, G. and Otto-Bliesner, B. 2001. Rapid (10-yr) recovery of terrestria productivity in a simulation study of the terminal Cretaceous impact event. *Earth and Planetary Science Letters* **192**, 137–144.

Lorenz, R. 2003. Full steam ahead—probably. *Science* **299**, 837–838.

Lorenz, R.D. 2002. Planets, life and the production of entropy. *International Journal of Astrobiology* **1**, 3–13.

Lorenz, R.D., Lunine, J.I., Withers, P.G. and McKay, C.P. 2001. Titan, Mars and Earth: entropy production by latitudinal heat transport. *Geophysical Research Letters* **28**, 415–418.

Lovelock, J. 1979. *Gaia*. Oxford: Oxford University Press.

Lovelock, J. 2000a. *The Ages of Gaia*, 2nd ed. Oxford: Oxford University Press.

Lovelock, J. 2000b. *Homage to Gaia*. Oxford: Oxford University Press.

Lovelock, J. 2003. The living Earth. *Nature* **426**, 769–770.

Lovelock J. 2006. *The Revenge of Gaia*. London: Allen Lane.

Lovelock, J. 2014. *A Rough Ride to the Future*. London: Allen Lane.

Lovelock, J.E. 1965. A physical basis for life detection experiments. *Nature* **207**, 568–570.

Lovelock, J.E. 1992. A numerical model for biodiversity. *Philosophical Transactions of the Royal Society B* **338**, 383–391.

Lovelock, J.E. and Margulis, L. 1974. Atmospheric homeostasis by and for the biosphere: the Gaia hypothesis. *Tellus* **16**, 2–10.

Lovelock, J.E. and Watson, A.J. 1982. The regulation of carbon dioxide and climate: Gaia or geochemistry. *Planet Space Science* **30**, 795–802.

Lovelock, J.E., Maggs, R.J. and Wade, R.J. 1973. Halogenated hydrocarbons in and over the Atlantic. *Nature* **241**, 194–196.

MacArthur, R.H. and Wilson, E.O. 1967. *The Theory of Island Biogeography*. Princeton: Princeton University Press.

MacMahon, J.A. 1985. *Deserts*. New York: Alfred A. Knopf.

Maddock, L. 1991. Effects of simple environmental feedback on some population models. *Tellus* **43B**, 331–337.

Madigan, M.T., Martinko, J.M., Stahl, D. and Clark, D. 2012. *Brock: Biology of Microorganisms*, 13th ed. Boston: Pearson.

Magri, D. 2008. Patterns of post-glacial spread and the extent of glacial refugia of European beech (*Fagus sylvatica*). *Journal of Biogeography* **35**, 450–463.

Maitland, P.S. and Campbell, R.N. 1992. *Freshwater Fishes*. London: Harper Collins.

Majerus, M.E.N. 2002. *Moths*. London: Harper Collins.

Majerus, M.E.N. (eds Roy, H.E. and Brown, P.M.J.) 2016. *A Natural History of Ladybird Beetles*. Cambridge: Cambridge University Press.

Margulis, L. 1971. The origin of plant and animal cells. *American Scientist* **59**, 230–235.

Margulis, L. and Chapman, M.J. 2009. *Kingdoms and Domains*. Amsterdam: Academic Press.

Margulis, L. and Lovelock, J.E. 1974. Biological modulation of the Earth's atmosphere. *Icarus* **21**, 471–489.

Margulis, L. and Sagan, D. 1995. *What Is Life?* London: Wiedenfeld and Nicolson.

Margulis, L. and Sagan, D. 1997. *Slanted Truths*. New York: Copernicus.

Marks, R.B. 2020. *The Origin of the Modern World*, 4th ed. Lanham: Rowman and Littlefield.

Martin, A.E., Lockhart, J.K. and Fahrig, L. 2023. Are weak dispersers more vulnerable than strong dispersers to land use intensification? *Proceedings of the Royal Society B* **290**, 20220909.

May, R.M. 1973. *Stability and Complexity in Model Ecosystems*. Princeton: Princeton University Press.

May, R.M. 1974. Biological populations with non-overlapping generations: stable points, stable cycles, and chaos. *Science* **186**, 645–647.

May, R.M. 1988. How many species are there on Earth? *Science* **241**, 1441–1449.

Maynard Smith, J. and Szathmáry, E. 1995. *The Major Transitions in Evolution*. Oxford: W.H. Freeman.

Mayr, E. 1982. *The Growth of Biological Thought*. Cambridge, MA: Harvard University Press.

Mayr, E. 2004. *What Makes Biology Unique?* Cambridge: Cambridge University Press.

Mayr, O. 1986. *Authority, Liberty and Automatic Machinery in Early Modern Europe*. Baltimore: Johns Hopkins University Press.

McCallum, H., Kikkawa, J. and Catterall, C. 2000. Density dependence in an island population of silvereyes. *Ecology Letters* **3**, 95–100.

McCarthy, T. and Rubidge, B. 2005. *The Story of Earth and Life: A Southern African Perspective on a 4.6-Billion-Year Journey*. Cape Town: Struik Nature.

Medawar, P. 1982. *Pluto's Republic*. Oxford: Oxford University Press.

Merilaita, S., Tuomi, J. and Jormalainen, V. 1999. Optimization of cryptic coloration in heterogeneous habitats. *Biological Journal of the Linnean Society* **67**, 151–161.

Merilaita, S., Lyytinen, A. and Mappes, J. 2001. Selection for cryptic coloration in a visually heterogeneous habitat. *Proceedings of the Royal Society B* **268**, 1925–1929.

Meshik, A.P. 2005. The workings of an ancient nuclear reactor. *Scientific American* **293**, 56–56.

Midgley, J., Lawes, M.J. and Chamaillé, S. 2010. Savanna woody plant dynamics: the role of fire and herbivory, separately and synergistically. *Australian Journal of Botany* **58**, 1–11.

Midgley, M. 1989. *Wisdom, Information and Wonder: What Is Knowledge for?* London: Routledge.

Midgley, M. 2001. *Science and Poetry*. London: Routledge.

Milcu, A., Lukac, M. and Ineson, P. 2012a. The role of closed ecological systems in carbon cycle models. *Climatic Change* **112**, 709–716.

Milcu, A., Lukac, M., Subke, J.-A., et al. 2012b. Biotic carbon feedbacks in a materially closed soil–vegetation–atmosphere system. *Nature Climate Change* **2**, 281–284.

Miller, J. 1978. *The Body in Question*. London: Jonathan Cape.

Miller, P.S. and Lacy, R.C. 2003. *VORTEX: A Stochastic Simulation of the Extinction Process. Version 9.21. User's Manual*. Apple Valley, MN: Conservation Specialist Group (SSC/IUCN).

Mills, D.B., Boyle, R.A., Daines, S.J., et al. 2022. Eukaryogenesis and oxygen in Earth history. *Nature Ecology and Evolution* **6**, 520–532.

Milne, F.I. and Abbott, R.J. 2000. Origin and evolution of invasive naturalized material of *Rhododendron ponticum* L. in the British Isles. *Molecular Ecology* **9**, 541–556.

Miskin, I., Rhodes, G., Lawlor, K., Saunders, J.R. and Pickup, R.W. 1998. Bacteria in post-glacial freshwater sediments. *Microbiology* **144**, 2427–2439.

Mitchell, R.L., Strullu-Derrien, C., Sykes, D., Pressel, S., Duckett, J.G. and Kenrick, P. 2020. Cryptogamic ground covers as analogues for early terrestrial biospheres: initiation and evolution of biologically mediated proto-soils. *Geobiology* **19**, 292–306.

Mitton, S. 2022. A short history of panspermia from Antiquity through the mid-1970s. *Astrobiology* **22**, 1379–1391.

Mo, Y., Peng, F., Gao, X., et al. 2021. Low shifts in salinity determined assembly processes and network stability of microeukaryotic plankton communities in a subtropical reservoir. *Microbiome* **9** (128).

Monteith, J.L. 1981. Evaporation and surface temperature. *Quarterly Journal of the Royal Meteorological Society* **107**, 1–27.

Monteith, J.L. and Unsworth, M.H. 1990. *Principles of Environmental Physics*, 2nd ed. London: Edward Arnold.

Mooney, H.A. and Ehleringer J.R. 1997. Photosynthesis. In: *Plant Ecology*, 2nd ed (ed. Crawley M.), pp 1–27. Oxford: Blackwell.

Moorbath, S. 2005. Dating earliest life. *Nature* **434**, 155.

Moore, P.D. 1975. Origin of blanket mires. *Nature* **256**, 267–269.

Moore, P.D. 1993. The origin of blanket mire, revisited. In: *Climate Change and Human Impact on the Landscape* (ed. Chambers, F.M.), pp 217–224. London: Chapman and Hall.

Moore, P.D. 2001. A never-ending story. *Nature* **409**, 565.

Moore, P.D. and Bellamy, D.J. 1974. *Peatlands*. London: Elek Science.

Morton, O. 2003. *Mapping Mars*. London: Forth Estate.

Morton, O. 2007. *Eating the Sun*. London: Forth Estate.

Moss, B. 2001. *The Broads*. London: Harper Collins.

Moss, B. 2012. *Liberation Ecology: The Reconciliation of Natural and Human Cultures*. Oldendorf (Luhe): International Ecology Institute.

Myers, N. and Kent, J. 2003. New consumers: the influence of affluence on the environment. *Proceedings of the National Academy of Sciences of the United States of America* **100**, 4963–4968.

Naeem, S. 1998. Species redundancy and ecosystem reliability. *Conservation Biology* **12**, 39–45.

Naeem, S., Thompson, L.J., Lawler, S.P., Lawton, J.H. and Woodfin, R.M. 1994. Declining biodiversity can alter the performance of ecosystems. *Nature* **368**, 734–737.

Narlikar, J.V. 1999. *Seven Wonders of the Cosmos*. Cambridge: Cambridge University Press.

Naudts, K., Chen, Y., McGrath, M.J., et al. 2016. Europe's forest management did not mitigate climate warming. *Science* **351**, 597–600.

Nee, S. 2007. Metapopulations and their dynamics. In: *Theoretical Ecology* (eds. May, R.M. and McLean, A.R.), pp 35–45. Oxford: Oxford University Press.

Needham, J. 1956. *Science and Civilisation in China. Vol 2*. Cambridge: Cambridge University Press.

Newmark, W.D. 1987. A land-bridge island perspective on mammalian extinctions in western North American parks. *Nature* **325**, 430–432.

Nicholson, A.E., Wilkinson, D,M., Williams, H.T.P. and Lenton, T.M. 2017. Multiple states of environmental regulation in well-mixed model biospheres. *Journal of Theoretical Biology* **414**, 17–34.

Nicholson, A.E., Wilkinson, D.M., Williams, H.T.P. and Lenton, T.M. 2018a. Gaian bottlenecks and planetary habitability maintained by evolving model biospheres: the ExoGaia model. *Monthly Notices of the Royal Astronomical Society* **477**, 727–740.

Nicholson, A.E., Wilkinson, D.M., Williams, H.T. and Lenton, T.M. 2018b. Alternative mechanisms for Gaia. *Journal of Theoretical Biology* **457**, 249–257.

Nicholson, W.L. 2009. Ancient micronauts: interplanetary transport of microbes by cosmic impacts. *Trends in Microbiology* **17**, 243–250.

Nisbet, E.G. 1991. *Leaving Eden: To Protect and Manage the Earth*. Cambridge: Cambridge University Press.

Nisbet, E.G. 2001. Where did early life live, and what was it like? In: *Earth Systems Science. A New Subject for Study (Geophysiology) or a New Philosophy*? (eds Guerzoni, S., Harding, S., Lenton, T. and Ricci Lucchi, F.), pp 43–52. Siena: International School Earth and Planetary Science.

Nisbet, E.G. and Fowler, C.M.R. 2003. The early history of life. *Treatise on Geochemistry* **8**, 1–39.

Nisbet, E.G. and Sleep, N.H. 2001. The habitat and nature of early life. *Nature* **409**, 1083–1091.

Nobis, M. and Wohlgemuth, T. 2004. Trend words in ecological core journals over the last 25 years (1978–2002). *Oikos* **106**, 411–421.

Norros, V., Holme, P., Norberg, A. and Ovaskainen, O. 2023. Spore production reveals contrasting seasonal stratergies and a trade-off between spore size and number of wood inhabiting fungi. *Functional Ecology* **37**, 551–563.

Novotny, V. and Basset, Y. 2005. Host specificity of insect herbivores in tropical forests. *Proceedings of the Royal Society B* **272**, 1083–1090.

Nunes-Neto, N.F., Do Carmo, R.S. and El-Hani, C.N. 2009. The relationship between marine phytoplankton, dimethylsulphide and the global climate: the CLAW hypothesis as a Lakatosian progressive problemshift. In: *Marine Phytoplankton* (eds Kersey, W.T. and Munger, S.P.), pp 1–17. Hauppauge, NY: Nova Science Publishers.

O'Brien, D.P., Lorenz, R.D. and Lunine, J.I. 2005. Numerical calculations on the longevity of impact oases on Titan. *Icarus* **173**, 243–253.

O'Malley, M.A. 2014. *Philosophy of Microorganisms*. Cambridge: Cambridge University Press.

O'Regan, H.J. and Turner, A. 2004. Biostratigraphic and palaeoecological implications of new fossil felid material from the Plio-Pleistocene site of Tegelen, The Netherlands. *Palaeontology* **47**, 1181–1193.

O'Regan, H.J., Turner, A. and Wilkinson, D.M. 2002. European Quaternary refugia: a factor in large carnivore extinction? *Journal of Quaternary Science* **17**, 789–795.

Odling-Smee, F.J., Laland, K.N. and Feldman, M.W. 1996. Niche construction. *American Naturalist* **147**, 641–648.

Odling-Smee, F.J., Laland, K.N. and Feldman, M.W. 2003. *Niche Construction*. Princeton: Princeton University Press.

Odum, E.P. 1969. The strategy of ecosystem development. *Science* **164**, 262–270.

Olby, R. 2009. *Francis Crick: Hunter of Life Secrets*. New York: Cold Spring Harbor Laboratory Press.

Osborne, A.W., Keighley, A.T., Ingleby, E.R., et al. 2021. From bare peat desert to nature reserve within ten years: a review of restoration practice on Little Woolden Moss, Manchester, UK. *North West Geography* **21**, 31–48.

Ozawa, H. and Ohmura, A. 1997. Thermodynamics of a global-mean state of the atmosphere—a state of maximum entropy increase. *Journal of Climate* **10**, 441–445.

Pace, N.R. 2001. The universal nature of biochemistry. *Proceedings of the National Acadamy of Sciences of the United States of America* **98**, 805–808.

Paltridge, G.W. 1975. Global dynamics and climate change: a system on minimum entropy exchange. *Quarterly Journal of the Royal Meteorological Society* **101**, 475–484.

Parker, A. 2005. *Seven Deadly Colours*. London: The Free Press.

Parker, A.R. 1998. Colour in Burgess Shale animals and the effect of light on evolution in the Cambrian. *Proceedings of the Royal Society B* **265**, 967–972.

Parks, S.A. and Harcourt, A.H. 2002. Reserve size, local human density, and mammalian extinctions in U.S. protected areas. *Conservation Biology* **16**, 800–808.

Patenaude, G.L., Briggs, B.D.J., Milne, R., Rowland, C.S., Dawson, T.P. and Pryor, S.N. 2003. The carbon pool in a British semi-natural woodland. *Forestry* **75**, 109–119.

Pausas, J.G. and Bond, W.J. 2018. Humboldt and the reinvention of nature. *Journal of Ecology* **107**, 1032–1037.

Pausas, J.G. and Bond W.J. 2022. Feedbacks in ecology and evolution. *Trends in Ecology and Evolution* **37**, 637–644.

Pedrós-Alió, C. 2012.The rare bacterial biosphere. *Annual Reviews Marine Science* **4**, 449–466.

Pegg, D.E. 2002. The history and principles of cryopreservation. *Seminars in Reproductive Medicine* **20**, 5–13.

Penesyan, A., Paulsen, I.T., Kjelleberg, S. and Gilling M.R. 2021. Three faces of biofilms: a microbial lifestyle, a nascent multicellular organism, and an incubator for biodiversity. *Biofilms and Microbiomes* **7**, 80.

Penrose, R. 2004. *The Road to Reality*. London: Jonathan Cape.

Perrin, J. 1990. *Yes to Dance*. Sparkford: Oxford Illustrated Press.

Perutz, M.F. 1987. Physics and the riddle of life. *Nature* **326**, 555–558.

Pielou, E.C. 2001. *The Energy of Nature*. Chicago: Chicago University Press.

Pimm, S.L. 1991. *The Balance of Nature?* Chicago: University of Chicago Press.

Pitelka, L.F. and the Plant Migration Workshop Group. 1997. Plant migration and climate change. *American Scientist* **85**, 464–473.

Polis, G.A. 1999. Why are parts of the world green? Multiple factors control productivity and the distribution of biomass. *Oikos* **86**, 3–15.

Pope, K.O., D'Hondt, S.L. and Marshall, C.R. 1998. Meteorite impact and the mass extinction of species at the Cretaceous/Tertiary boundary. *Proceedings of the National Academy of Sciences of the United States of America* **95**, 11028–11029.

Porada, P., Weber, B., Elbert, W., Pöschl, U. and Kleidon, A. 2014. Estimating impacts of lichens and bryophytes on global biogeochemical cycles. *Global Biogeochemical Cycles* **28**, 71–85.

Porley, R. and Hodgetts, N. 2005. *Mosses and Liverworts*. London: Harper Collins.

Porter, R. 1997. *The Greatest Benefit to Mankind*. London: Harper Collins.

Porter, R. 2000. *Enlightenment*. London: Allen Lane.

Postgate, J. 1990. The microbes that would not die. *New Scientist* **127** (1726), 46–49.

Pounds, J.A., Bustamante, M.R., Coloma, L.A., et al. 2006. Widespread amphibian extinctions from epidemic disease driven by global warming. *Nature* **439**, 161–167.

Preston, C.D., Pearman, D.A. and Dines, T.D. 2002. *New Atlas of the British and Irish Flora*. Oxford: Oxford University Press.

Pringle, R.M. 2020. Untangling food webs. In: *Unsolved Problems in Ecology* (eds Dobson, A., Holt, R.D. and Tilman, D.), pp 225–238. Princeton: Princeton University Press.

Purvis, A. and Hector, A. 2000. Getting the measure of biodiversity. *Nature* **405**, 212–219.

Qiu, C., Ciais, P., Zhu, D., et al. 2021. Large historical carbon emissions from cultivated northern peatlands. *Science Advances* **7**, eabf1332.

Quammen, D. 2018. *The Tangled Tree*. London: Collins.

Rainey, P.B. and Travisano, M. 1998. Adaptive radiation in a heterogenous environment. *Nature* **394**, 69–72.

Ramsbottom, J. 1953. *Mushrooms and Toadstools*. London: Collins.

Ratcliffe, N., Pelembe, T. and White, R. 2008. Resolving the population status of Ascension frigatebird *Fregata aquila* using a 'virtual ecology' model. *Ibis* **150**, 300–306.

Raven, J.A. 1998. The twelfth Tansley lecture. Small is beautiful: the picophytoplankton. *Functional Ecology* **12**, 503–513.

Raven, P.H. and Johnson, G.B. 1999. *Biology*, 5th ed. Boston: McGraw-Hill.

Redfield, A.C. 1958. The biological control of the chemical factors in the environment. *American Scientist* **46**, 205–221.

Reid, C. 1899. *The Origin of the British Flora*. London: Dulau and Co.

Renne, P.R., Deino, A.L., Hilgen, F.J., et al. 2013. Time scales of critical events around the Cretaceous–Paleogene boundary. *Science* **339**, 684–687.

Retallack, G.J. 1997. Early forest soils and their role in Devonian global change. *Science* **276**, 583–585.

Retallack, G.J. 2000. Ordovician life on land and early Palaeozoic global change. *The Palaeontological Society Papers* **6**, 21–45.

Retallack, G.J. 2001. *Soils of the Past*, 2nd ed. Oxford: Blackwell Science.

Ricklefs, R.E. 1976. *The Economy of Nature*. Portland: Chiron Press.

Ricklefs, R.E. and Relyea, R. 2014. *Ecology*, 7th ed. New York: W.H. Freeman.

Ridgwell, A. and Zeebe, R.E. 2005. The role of the global carbonate cycle in the regulation and evolution of the Earth system. *Earth and Planetary Science Letters* **234**, 299–315.

Ridley, H.N. 1930. *The Dispersal of Plants Throughout the World*. Ashford: L. Reeve and Co.

Riebesell, U., Zondervan, I., Rost, B., Tortell, P.D., Zeebe, R.E. and Morel, F.M.M. 2000. Reduced calcification of marine plankton in response to increased atmospheric CO_2. *Nature* **407**, 364–367.

Rigby, S. and Milsom, C.V. 2000. Origins, evolution, and diversification of zooplankton. *Annual Reviews of Ecology and Systematics* **31**, 293–313.

Robinson, J.M. 1990. Lignin, land plants, and fungi: biological evolution affecting Phanerozoic oxygen balance. *Geology* **15**, 607–610.

Robinson, J.M. 1991. Fire in Phanerozoic cybernetics. In: *Scientists on Gaia* (eds Schneider, S.H. and Boston, P.J.), pp 363–372. Cambridge, MA: MIT Press.

Rocap, G., Larimer, F.W., Lameridin, J., et al. 2003. Genome divergence in two *Prochlorococcus* ecotypes reflects oceanic niche differentiation. *Nature* **424**, 1042–1047.

Rodwell, J.S. (ed.) 1991. *British Plant Communities Vol 2. Mires and Heaths*. Cambridge: Cambridge University Press.

Rosell, F., Bozsér, O., Collen, P. and Parker, H. 2005. Ecological impact of beavers *Castor fiber* and *Castor canadensis* and their ability to modify ecosystems. *Mammal Review* **35**, 248–276.

Rosing, M.T. 1999. ^{13}C-depleted carbon microparticles in >3700-Ma sea-floor sedimentary rocks from West Greenland. *Science* **283**, 674–676.

Rousk, K. 2022. Biotic and abiotic controls of nitrogen fixation in cyanobacteria–moss associations. *New Phytologist* **235**, 1330–1335.

Rowantree, R.A. and Nowak, D.J. 1991. Quantifying the role of urban forests in removing atmospheric carbon dioxide. *Journal of Arboriculture* **17**, 269–275.

Rubin, S. and Crucifix, M. 2022. Taking the Gaia hypothesis at face value. *Ecological Complexity* **49**, 100981.

Rubin, S., Parr, T., Costa, L.D. and Friston, K. 2020. Future climates: Markov blankets and active inference in the biosphere. *Journal Royal Society Interface* **17**, 20200503.

Ruddiman, W.F. 2003. The anthropic greenhouse era began thousands of years ago. *Climatic Change* **61**, 261–293.

Ruddiman, W.F. 2005. *Plows, Plagues and Petroleum*. Princeton: Princeton University Press.

Ruddiman, W.F., He, F., Vavrus, S.J. and Kutzbach, J.E. 2020. The early anthropic hypothesis: a review. *Quaternary Science Reviews* **240**, 106386.

Ruxton, G.D., Allen, W.L., Sherratt, T.N. and Speed, M.P. 2018. *Avoiding Attack: The Evolutionary Ecology of Crypsis, Aposematism, and Mimicry*, 2nd ed. Oxford: Oxford University Press.

Rydin, H. and Jeglum, J.K. 2013. *The Biology of Peatlands*, 2nd ed. Oxford: Oxford University Press.

Sagan, C. and Chyba, C. 1997. The early faint sun paradox: organic shielding of ultraviolet-labile greenhouse gasses. *Science* **276**, 1217–1221.

Sagan, C., Thompson, W.R., Carlson, R., Gurnett, D. and Hord, C. 1993. A search for life on Earth from the Galileo spacecraft. *Nature* **365**, 715–721.

Santini, F. and Galleni, L. 2001. Non-light based ecosystems and bioastronomy. *Rivista di Biologia* **94**, 427–442.

Sapp, J. 1994. *Evolution by Association: A History of Symbiosis*. Oxford: Oxford University Press.

Sarkar, S. 2013. Erwin Schrödinger's excursus on genetics. In: *Outsider Scientists: Routes to Innovation in Biology* (eds Harman, O. and Dietrch, M.R.), pp 93–109. Chicago: University of Chicago Press.

Saunders, P.T. 1994. Evolution without natural selection: further implications of the Daisyworld parable. *Journal of Theoretical Biology* **166**, 365–373.

Schama, S. 2000. *A History of Britain Vol 1. 300BC–1603 AD*. London: BBC.

Schimper, A.F.W. 1903. *Plant-Geography Upon a Physiological Basis*. Oxford: The Clarendon Press. [English translation of a book first published in German in 1898.]

Schopf, J.W. 1999. *Cradle of Life, the Discovery of Earth's Earliest Fossils*. Princeton: Princeton University Press.

Schrödinger, E. 1948. *What Is Life?* Cambridge: Cambridge University Press.

Schwartzman, D. 1999. *Life, Temperature, and the Earth*. New York: Columbia University Press.

Schwartzman, D.W. and Volk, T. 1989. Biotic enhancement of weathering and the habitability of Earth. *Nature* **340**, 457–460.

Searle, C.L. and Christie, M.R. 2021. Evolutionary rescue in host pathogen systems. *Evolution* **75**, 2948–2958.

Seaward, M.R.D. and Pentecost, A. 2001. Lichen flora of the Malham Tarn area. *Field Studies* **10**, 57–92.

Sheldrake, M. 2020. *Entangled Life*. London: The Bodley Head.

Sherratt, T.N. and Wilkinson, D.M. 2009. *Big Questions in Ecology and Evolution*. Oxford: Oxford University Press.

Sherratt, T.N., Pollitt, D. and Wilkinson, D.M. 2007. The evolution of crypsis in replicating populations of web-based prey. *Oikos* **116**, 449–460.

Shklovskii, I.S. and Sagan, C. 1977. *Intelligent Life in the Universe*. London: Picador.

Shorrocks, B. and Bates, W. 2015. *The Biology of African Savannahs*, 2nd ed. Oxford: Oxford University Press.

Silvertown, J. 2005. *Demons in Eden*. Chicago: University of Chicago Press.

Silvertown, J. 2015. Have ecosystem services been oversold? *Trends in Ecology and Evolution* **30**, 641–647.

Simberloff, D. 1998a. Small and declining populations. In: *Conservation Science and Action* (ed. Sutherland, W.J.), pp 116–134. Oxford: Blackwell.

Simberloff, D. 1998b. Flagships, umbrellas, and keystones: is single-species management passé in the landscape era? *Biological Conservation* **83**, 247–257.

Simó, R. 2001. Production of atmospheric sulfur by oceanic plankton: biogeochemical, ecological and evolutionary links. *Trends in Ecology and Evolution* **16**, 287–294.

Sinclair, A.R.E. 1989. Population regulation in animals. In: *Ecological Concepts* (ed. Cherratt, J.M.), pp 197–241. Oxford: Blackwell.

Slack, N.G. 2010. *G. Evelyn Hutchinson and the Invention of Modern Ecology*. New Haven: Yale University Press.

Sleep, N., Zahnle, K.J., Kasting, J.F. and Marowitz, H.J. 1989. Annihilation of ecosystems by large asteroid impacts on early Earth. *Nature* **342**, 139–142.

Slobodkin, L.B. and Slack, N.G. 1999. George Evelyn Hutchinson: 20th-century ecologist. *Endeavour* **23**, 24–30.

Smith, M.D., Koerner, S.E., Knapp, A.K., et al. 2020. Mass ratio effects underlie ecosystem response to environmental change. *Journal of Ecology* **108**, 855–864.

Smolin, L. 1997. *The Life of the Cosmos*. London: Weidenfeld and Nicolson.

Snow, C.P. 1959. *The Two Cultures*. Cambridge: Cambridge University Press.

Sommer, C. and Bibby, B.M. 2002. The influence of veterinary medicines on the decomposition of dung organic matter in soil. *European Journal of Soil Biology* **38**, 155–159.

Song, X., Xu, H. and Bai, Y. 2022. The systematic out-branching (dragon style) rivers under the perspective of connection between river morphology and ecology. *Ecohydrology & Hydrobiology* **22**, 505–510.

Sorensen, R. 1991. Thought experiments. *American Scientist* **79**, 250–263.

Spooner, B. and Roberts, P. 2005. *Fungi*. London: Collins.

Spudis, P.D., Wilhelms, D.E. and Robinson, M.S. 2011. The sculptured hills of the Taurus highlands: implications for the relative age of Serenitatis basin chronologies. *Journal of Geophysical Research: Planets* **116**, E00H03.

Stace, C. 2019. *New Flora of the British Isles*, 4th ed. Middlewood Green: C&M Floristics.

Stace, C.A. and Crawley, M.J. 2015. *Alien Plants*. London: William Collins.

Steffen, W., Richardson, K., Rockström, J., et al. 2020. The emergence and evolution of Earth system science. *Nature Reviews Earth and Environment* **1**, 54–63.

Stephens, P.A. and Sutherland, W.J. 1999. Consequences of the Allee effect for behaviour, ecology and conservation. *Trends in Ecology and Evolution* **14**, 401–405.

Stewart, I. 1998. *Life's Other Secret*. London: Allen Lane.

Stewart, I. 2012. *Seventeen Equations that Changed the World*. Profile Books: London.

Stork, N. E. 2018. How many species of insects and other terrestrial arthropods are there on Earth? *Annual Review of Entomology* **63**, 31–45.

Sugimoto, T. 2002. Darwinian evolution does not rule out the Gaia hypothesis. *Journal of Theoretical Biology* **218**, 447–455.

Sullivan, M.B., Waterbury, J.B. and Chisholm, S.W. 2003. Cyanophages infecting the oceanic cyanobacterium Prochlorococcus. *Nature* **424**, 1047–1051.

Sunda, W., Kieber, D.J., Kiene, R.P. and Huntsman, S. 2002. An antioxidant function for DMSP and DMS in marine algae. *Nature* **418**, 317–320.

Suttle, C.A. 2005. Viruses in the sea. *Nature* **437**, 356–361.

Sutton, G., Bennett, J. and Bateman, M. 2014. Effects of ivermectin residues on dung invertebrate communities in a UK farmland habitat. *Insect Conservation and Diversity* **7**, 64–72.

Tansley, A.G. 1920. The classification of vegetation and the concept of development. *Journal of Ecology* **8**, 118–144.

Tansley, A.G. 1935. The use and abuse of vegetational concepts and terms. *Ecology* **16**, 284–307.

Tansley, A.G. 1949. *Britain's Green Mantle*. London: George Allen and Unwin.

Terhaar, J., Frölicher, T.L. and Joos, F. 2022. Observation-constrained estimates of the global ocean carbon sink for Earth system models. *Biogeosciences* **19**, 4431–4457.

Thomas, C.D. 2017. *Inheritors of the Earth*. London: Allen Lane.

Thorne, K.S. and Blandford, R.D. 2017. *Modern Classical Physics*. Princeton: Princeton University Press.

Tice, M.M. and Lowe, D.R. 2004. Photosynthetic microbial mats in the 3416-Myr-old ocean. *Nature* **431**, 549–552.

Tickell, C. 2011. Societal responses to the Anthropocene. *Philosophical Transactions of the Royal Society A* **369**, 926–932.

Tilman, D. 2000. Causes, consequences and ethics of biodiversity. *Nature* **405**, 208–211.

Tilman, D., Lehman, C.L. and Bristow, C.E. 1998. Diversity–stability relationships: statistical inevitability or ecological consequence? *American Naturalist* **151**, 277–282.

Tilman, D., Wedin, D. and Knops, J. 1996. Productivity and sustainability influenced by biodiversity in grassland ecosystems. *Nature* **379**, 718–720.

Tolonen, K., Warner, B.G. and Vasander, H. 1992. Ecology of Testaceans (Protozoa: Rhizopoda) in mires in southern Finland: 1. Autecology. *Archiv fur Protistenkunde* **142**, 119–138.

Tompkins, D.M. and Begon, M. 1999. Parasites can regulate wildlife populations. *Parisitology Today* **15**, 311–312.

Tyrrell, T. 1999. The relative influence of nitrogen and phosphorus on oceanic primary production. *Nature* **400**, 525–531.

Tyrrell, T. 2004. Biotic plunder: control of the environment by biological exhaustion of resources. In: *Scientists Debate Gaia* (eds Schneider, S.H., Miller, J.R.,Crist, E. and Boston, P.J.), pp 137–147. Cambridge, MA: MIT Press.

Tyrrell, T. 2013. *On Gaia*. Princeton: Princeton University Press.

Tyrrell, T. 2020. Chance played a role in determining whether Earth stayed habitable. *Communications Earth & Environment* **1** (61).

Van Damme, R., Wilson, R.S., Vanhooydonck, B. and Aerts, P. 2002. Performance constraints in decathletes. *Nature* **415**, 755–756.

Van der Pijl, L. 1969. *Principles of Dispersal in Higher Plants*. Berlin: Springer.

Van der Valk, AG. 2014. From formation to ecosystem: Tansley's response to Clements' climax. *Journal of the History of Biology* **47**, 293–321.

Vepsäläinen, K. and Spence, J.R. 2000. Generalization in ecology and evolutionary biology: from hypothesis to paradigm. *Biology and Philosophy* **15**, 211–238.

Vernadsky, V.I. 1998. *The Biosphere*. New York: Copernicus. [First published 1926.]

Vincent, W.F. 2000. Cyanobacterial dominance in the polar regions. In: *The Ecology of Cyanobacteria* (eds Whitton, B.A. and Potts, M.), pp 321–340. Dordrecht: Kluwer.

Vitali, R., Belcher, C.M., Kaplan, J. and Watson, A.J. 2022. Increased fire activity under high atmospheric oxygen concentrations is compatable with the presence of forests. *Nature Communications* **13**, 7285.

Vitousek, P. 2004. *Nutrient Cycling and Limitation*. Oxford: Princeton University Press.

Volk, T. 1987. Feedbacks between weathering and atmospheric CO_2 over the last 100 million years. *American Journal of Science* **287**, 763–779.

Volk, T. 1998. *Gaia's Body*. New York: Copernicus.

Volk, T. 2004. Gaia is life in a wasteworld of by-products. In: *Scientists Debate Gaia* (eds Schneider, S.H., Miller, J.R., Crist, E. and Boston, P.J.), pp 27–36. Cambridge, MA: MIT Press.

Volk, T. 2008. *CO_2 Rising*. Cambridge, MA: MIT Press.

Vreeland, R.H., Rosenzweig, W.D. and Powers, D.W. 2000. Isolation of a 250 million-year-old halotolerant bacterium from a primary salt crystal. *Nature* **407**, 897–900.

Walker, J.C.G., Hays, P.B. and Kasting, J.F. 1981. A negative feedback mechanism for the long-term stabilisation of Earth's surface temperature. *Journal of Geophysical Research* **86**, 9776–9782.

Wall, R. and Strong, L. 1987. Environmental consequences of treating cattle with the antiparasitic drug ivermectin. *Nature* **327**, 418–421.

Waltham, D. 2014. *Lucky Planet*. London: Icon Books.

Wang, J., We, Q., Liu, J., et al. 2019. Vehicle emission and atmospheric pollution in China: problems, progress, and prospects. *PeerJ* **7**, e6932.

Warming, E. 1909. *Oecology of Plants*. Oxford: The Clarendon Press. [English edition of a book first published in Danish in 1895.]

Watson, A., Lovelock, J.E. and Margulis, L. 1978. Methanogenesis, fires and the regulation of atmospheric oxygen. *BioSystems* **10**, 293–298.

Watson, A.J. 1999. Coevolution of the Earth's environment and life: Goldilocks, Gaia and the anthropic principle. In: *James Hutton—Present and Future* (eds Craig, G.Y. and Hull, J.H.), pp 75–88. Geological Society Special Publication 150. London: Geological Society.

Watson, A.J. 2004. Gaia and observer self selection. In: *Scientists Debate Gaia* (eds Schneider, S.H., Miller, J.R., Crist, E. and Boston, P.J.), pp 201–208. Cambridge, MA: MIT Press.

Watson, A.J. and Lovelock, J.E. 1983. Biological homeostasis of the global environment: the parable of Daisyworld. *Tellus* **35B**, 284–289.

Watson, A.J., Bakker, D.C.E., Ridgewell, A.J., Boyd, P.W. and Law, C.S. 2000. Effect of iron supply on Southern Ocean CO_2 uptake and implications for glacial atmospheric CO_2. *Nature* **407**, 730–733.

Watson, R.A. and Szathmáry, E. 2016. How can evolution learn? *Trends in Ecology and Evolution* **31**, 147–157.

Wells, H.G., Huxley, J. and Wells, G.P. 1931. *The Science of Life*. Vol. 3. London: The Waverley Book Company.

Wells, L.E., Armstrong, J.C. and Gonzalez, G. 2003. Reseeding of early Earth by impacts of returning ejecta during the late heavy bombardment. *Icarus* **162**, 38–46.

Welsh, D.T. 2000. Ecological significance of compatible solute accumulation by micro-organisms: from single cells to global climate. *FEMS Microbiology Reviews* **24**, 263–290.

West, S.A., Lively, C.M. and Read, A.F. 1999. A pluralist approach to sex and recombination. *Journal of Evolutionary Biology* **12**, 1003–1012.

Wheldale, M. 1916. *The Anthocyanin Pigments of Plants*. Cambridge: Cambridge University Press.

Whitman, W.B., Coleman, D.C. and Wiebe, W.J. 1998. Prokaryotes: the unseen majority. *Proceedings of the National Academy of Sciences of the United States of America* **95**, 6578–6583.

Whitton, B.A. 2012. *Ecology of Cyanobacteria II*. Dordrecht: Springer.

Wildman Jr, R.A., Hickey, L.J., Dickinson, M.B., et al. 2004. Burning of forest materials under late Paleozoic high atmospheric oxygen levels. *Geology* **32**, 457–460.

Wilkinson, D.M. 1997. Plant colonization: are wind dispersed seeds really dispersed by birds at larger spatial and temporal scales? *Journal of Biogeography* **24**, 61–65.

Wilkinson, D.M. 1998. Fragments of an entangled bank: do ecologists study most of ecology? *Oikos* **84**, 533–536.

Wilkinson, D.M. 1999. Is Gaia conventional ecology? *Oikos* **84**, 533–536.

Wilkinson, D.M. 2000. Running with the Red Queen: reflections on 'sex versus non-sex versus parasite'. *Oikos* **91** 589–596.

Wilkinson, D.M. 2001a. What is the upper size limit for cosmopolitan distribution in free-living microorganisms? *Journal of Biogeography* **28**, 285–291.

Wilkinson, D.M. 2001b. At cross purposes. *Nature* **412**, 485.

Wilkinson, D.M. 2003. The fundamental processes in ecology: a thought experiment on extraterrestrial biospheres. *Biological Reviews* **78**, 171–179.

Wilkinson, D.M. 2004a. The parable of Green mountain: Ascension Island, ecosystem construction and ecological fitting. *Journal of Biogeography* **31**, 1–4.

Wilkinson, D.M. 2004b. Homeostatic Gaia: an ecologist's perspective on the possibility of regulation. In: *Scientists Debate Gaia* (eds Schneider, S.H., Miller, J.R., Crist, E. and Boston, P.J.), pp 71–76. Cambridge, MA: MIT Press.

Wilkinson, D.M. 2006. *Fundamental Processes in Ecology: An Earth Systems Approach*. Oxford: Oxford University Press.

Wilkinson, D.M. 2015. On Gaia: a critical investigation of the relationship between life and Earth. *International Journal of Environmental Studies* **72**, 724–730.

Wilkinson, D.M. 2021a. *Ecology and Natural History*. London: Collins.

Wilkinson, D.M. 2021b. Peering into the future with help from the past: the importance of long-term ecological studies. *British Wildlife* **32**, 556–562.

Wilkinson, D.M. 2022. Testate amoebae—a beginner's guide to a fascinating microworld. *British Wildlife* **34**, 96–104.

Wilkinson, D.M. and Davis, S.R. 2005. An introduction to Chat Moss. In: *Quaternary of Rossendale Forest and Greater Manchester* (ed. Croft, R.G.), pp 38–43. London: Quaternary Research Association.

Wilkinson, D.M. and O'Regan, H.J. 2003. Modelling differential extinctions to understand big cat distribution on Indonesian islands. *Global Ecology and Biogeography* **12**, 519–524.

Wilkinson, D.M. and Sherratt, T.N. 2001. Horizontally acquired mutualisms, an unsolved problem in ecology? *Oikos* **92**, 377–384.

Wilkinson, D.M. and Sherratt, T.N. 2016. Why is the world green? The interactions of top-down and bottom-up processes in terrestrial vegetation ecology. *Plant Ecology and Diversity* **9**, 127–140.

Wilkinson, D.M., Sherratt, T.N., Phillip, D.M., Wratten, S.D., Dixon, A.G.F. and Young, A.J. 2002. The adaptive significance of autumn leaf colours. *Oikos* **99**, 402–407.

Wilkinson, D.M., Koumoustsaris, S., Mitchell, E.A.D. and Bey, I. 2012. Modelling the effect of size on the aerial dispersal of microorganisms. *Journal of Biogeography* **39**, 89–97.

Wilkinson, D.M., Midgley, J.J. and Cunningham, A.B. 2022. Constraints crashes and conservation: were historical African savanna elephant *Loxodonta africana* densities relatively high or lower than those seen in protected areas today? *Plant Ecology and Diversity* **15**, 1–11.

Williams, G.C. 1992a. Gaia, nature worship and biocentric fallacies. *Quarterly Review of Biology* **67**, 479–486.

Williams, G.C. 1992b. *Natural Selection: Domains, Levels, and Challenges*. Oxford: Oxford University Press.

Williams, H.T.P. and Lenton, T.M. 2007. The Flask model: emergence of nutrient-recycling microbial ecosystems and their disruption by environment-altering 'rebel' organisms. *Oikos* **116**, 1087–1105.

Williams H.T.P. and Lenton, T.M. 2008. Environmental regulation in a network of simulated microbial ecosystems. *Proceedings of the National Academy of Sciences of the United States of America* **105**, 10432–10437.

Willis, K.J., Gillson, L. and Brncic, T.M. 2004. How 'virgin' is virgin rainforest? *Science* **304**, 402–403.

Wilson, E.O. 1992. *The Diversity of Life*. Cambridge, MA: Harvard University Press.

Wilson, K., Law, A., Gaywood, M., Ramsay, P. and Willby, N. 2020. Beavers: the original engineers of Britain's freshwaters. *British Wildlife* **31**, 403–411.

Wolstencroft, R.D. and Raven, J.A. 2002. Photosynthesis: likelihood of occurrence and possible detection on Earth-like planets. *Icarus* **157**, 535–548.

Wood, A.J., Ackland, G.J., Dyke, J.G., Williams, H.T.P. and Lenton, T.M. 2008. Daisyworld: a review. *Reviews of Geophysics* **46**, RG1001.

Wynn-Williams, D.D. 1996. Antarctic microbial diversity: the basis of polar ecosystem processes. *Biodiversity and Conservation* **5**, 1271–1293.

Xiong, J., Fischer, W.M., Inoue, K., Nakahara, M. and Bauer, C.E. 2000. Molecular evidence for the early evolution of photosynthesis. *Science* **289**, 1724–1730.

Yang, J., Smith, H.G., Sherratt, T.N. and Wilkinson, D.M. 2010. Is there a size limit for cosmopolitan distribution in free-living microorganisms? A biogeographical analysis of testate amoebae from polar areas. *Microbial Ecology* **59**, 635–645.

Yu, D.W. 2001. Parasites of mutualisms. *Biological Journal of the Linnean Society* **72**, 529–546.

Zambell, C.B., Adams, J.M., Goring, M.L. and Schwartzman, D.W. 2012. Effect of lichen colonisation on chemical weathering of hornblende granite as estimated by aqueous elemental flux. *Chemical Geology* **291**, 166–174.

Zellner, N.E.B. 2017. Cataclysm no more: new views on the timing and delivery of lunar impactors. *Origin of Life and Evolution of Biospheres* **47**, 261–280.

Zhang, B., DeAngelis, D.L. and Wei-Ming, N. 2021. Carrying capacity of spatially distributed metapopulations. *Trends in Ecology and Evolution* **36**, 164–173.

Zhang, X., Shu, D., Li, Y. and Han, J. 2001. New sites of Chengjiang fossils: crucial windows on the Cambrian explosion. *Journal of the Geological Society, London* **158**, 211–218.

Ziman, J. 2003. Emerging out of nature into history: the plurality of the sciences. *Philosophical Transactions of the Royal Society London A* **361**, 1617–1633.

Index

People mentioned in the main text are indexed but not those only cited in references.